WORKED EXAMPLES IN X-RAY ANALYSIS

PHILIPS TECHNICAL LIBRARY

WORKED EXAMPLES IN X-RAY ANALYSIS

R. JENKINS

J.L. DE VRIES

Springer Science+Business Media, LLC

Originally published by Springer-Verlag New York Inc. in 1970
Softcover reprint of the hardcover 1st edition 1970

First edition 1970
Reprinted 1971
Reprinted 1972

Additional material to this book can be downloaded from http://extras.springer.com

ISBN 978-1-4899-2649-4 ISBN 978-1-4899-2647-0 (eBook)
DOI 10.1007/978-1-4899-2647-0

PHILIPS

Preface

The purpose of this book is to provide the reader with a series of worked examples in X-ray spectrometry and X-ray diffractometry, in such a way that each example can be treated either as a posed question, i.e. one to which the reader must himself provide an answer, or as a guide to the method of treating a similar problem. The latter, of course, also provides a check on the answer produced by the reader. Although much basic theory can be derived by study of this work the first intention of the book is not to provide a source of basic theoretical knowledge in X-ray analysis. It is hoped that the book will be utilized more as a guide line in the tackling of theoretical and practical problems and as a means of establishing whether or not the reader is able to work out for himself a certain type of problem. For example, the series of examples on counting statistics has been chosen in such a way that after working through and understanding these, the reader should be able to handle most of the calculations that he is likely to come up against in this area.

Our own experience of most of the textbooks in this field is that although detailed explanation is given of the reason why a typical problem occurs and what can be done to overcome it, the reader is often unable to judge whether or not he is really able to apply this newly found knowledge to one of his own problems.

We look upon this book, therefore, as a supplement to existing work in the field rather than an independent source of information directly usable by a complete newcomer. The work as such covers both X-ray spectrometry and X-ray powder diffraction and some basic theory in both techniques is required before the problems can be attempted. All of the necessary theory in X-ray spectrometry will be found in our previous volume on "Practical X-Ray Spectrometry"[1] and a similar volume on "Practical X-Ray Diffractometry" is currently in preparation. Since the publication of the present work is likely to precede "Practical X-Ray Diffractometry" by a considerable time margin we would recommend the excellent books by Klug and Alexander[2] and Parrish[3] as good alternatives.

All of the worked examples have been divided into five major sections dealing with Spectra, Instrumentation, Counting Statistics, Quantitative Analysis and Miscellaneous. Within each of the five sections examples of three types will be found i.e. either of a general nature or dealing specifically with X-ray spectrometry or X-ray diffractometry. It is thus the intention that examples falling within the general category be used either for spectrometry or for diffractometry. Generally speaking, within a given definition (e.g. quantitative analysis/X-ray spectrometry) the difficulty of the examples increases with the number of the question. The exception to this is the miscellaneous section which, as its title would suggest, contains a number of examples of an undefined nature considered necessary to give a more equal coverage of typical practical problems.

In order to aid the reader in establishing how difficult an example is likely to prove, we have chosen to give each example a "difficulty index" denoted by A, B or C. A indicates a simple problem that should give little difficulty even to a relative newcomer to the field. B indicates a more difficult problem which requires a fair degree of theoretical knowledge but even so, should not present too much difficulty to somebody working in the field. C on the other hand indicates quite a difficult problem, many of these being in the form of unexpected results from well established experiments, requiring explanation rather than mathematical derivation.

Most of the worked examples cited in this book are based on problems which we ourselves have come up against and most of them have been utilized in some form or another in one of our own schools or symposia.

Some of the examples have been designed and checked by assistants in our laboratories and we are particularly grateful to Mr Böinck, Mr Hertroys and Mr Koekenbier for their help in this direction. We would also like to thank Janine de Brouwer for her considerable patience in the typing of the manuscript.

R. Jenkins

J.L. de Vries

Contents

Q19	Trace analysis - derivation of detection limit expression	Spectrometry	B
Q20	Choice between fixed time and fixed count	Spectrometry	B
Q21	Choice between ratio and absolute counting methods	Spectrometry	B
Q22	Variation of limit of detection with analysis time	Spectrometry	A
Q23	Analytical error due to counting statistics	Diffractometry	A
Q24	Particle statistics in X-ray diffractometry	Diffractometry	C

QUANTITATIVE ANALYSIS

Q25	Determination of traces of lead in oil (absolute measurements)	Spectrometry	A
Q26	Determination of CaO in cement (ratio measurements)	Spectrometry	A
Q27	Determination of nickel in steel (with dead time correction)	Spectrometry	A
Q28	Calculation of matrix absorption and prediction of absorption effects	Spectrometry	B
Q29	Calculation of absorption coefficients and absorption effects	Spectrometry	B
Q30	Absorption correction involving primary and secondary absorption	Spectrometry	B
Q31	Calculation of quantity of heavy absorber needed to minimize matrix effects	Spectrometry	B
Q32	Use of a correction factor (intensity correction)	Spectrometry	B
Q33	Use of a correction factor (concentration correction)	Spectrometry	B
Q34	Excitation probability	Spectrometry	C
Q35	Quantitative diffractometry - determination of Anatase in Rutile	Diffractometry	A

Q36	Quantitative diffractometry - use of internal standard	Diffractometry	B
Q37	Estimation of expected intensities in mixtures	Diffractometry	B
Q38	Quantitative X-ray diffractometry	Diffractometry	C

MISCELLANEOUS

Q39	Determination of dead time	General	B
Q40	Estimation of gold plating thickness	General	B
Q41	Measurement of coating thickness on irregular shaped objects		C
Q42	Background variation	Spectrometry	C
Q43	Use of primary filters for removal of characteristic tube lines	Speçtrometry	A
Q44	Calculation of β filter thickness and transmission	Diffractometry	A
Q45	Utilization of the β-filter and pulse height selector in diffractometry	Diffractometry	C
Q46	Determination of lattice constants for a cubic lattice	Diffractometry	B
Q47	Line broadening	Diffractometry	B

QUESTION 1

Characteristic spectra   Moseley's law

Moseley's law relates the wavelength λ of a characteristic line in a given series with atomic number Z. The usual expression of the law is

$$\frac{1}{\sqrt{\lambda}} = K(Z - \sigma) \qquad (1)$$

where K is a constant for a given spectral series and σ is a shielding constant.

A Using the data given, plot $1/\sqrt{\lambda}$ against Z to verify the law.

B Derive values for the constants Z and σ; then use Moseley's equation to estimate the $K\alpha$ wavelength of maganese ($Z = 25$).

Z	element	λ Å ($K\alpha$)
12	Mg	9.888
15	P	6.155
17	Cl	4.729
20	Ca	3.360
22	Ti	2.750
24	Cr	2.291
26	Fe	1.937

ANSWER 1

A Rearranging the given form of Moseley's equation one obtains

$$Z = \frac{1}{K\sqrt{\lambda}} + \sigma$$

Thus by plotting Z on the ordinate and $1/\sqrt{\lambda}$ on the abscissa, a curve will be obtained of slope = $1/K$ with an intercept of σ on the ordinate. Fig.1 shows such a plot using the given data, i.e.

Z	λ	$1/\sqrt{\lambda}$
12	9.888	0.3180
15	6.155	0.4031
17	4.729	0.4598
20	3.360	0.5455
22	2.750	0.6030
24	2.291	0.6607
26	1.937	0.7185

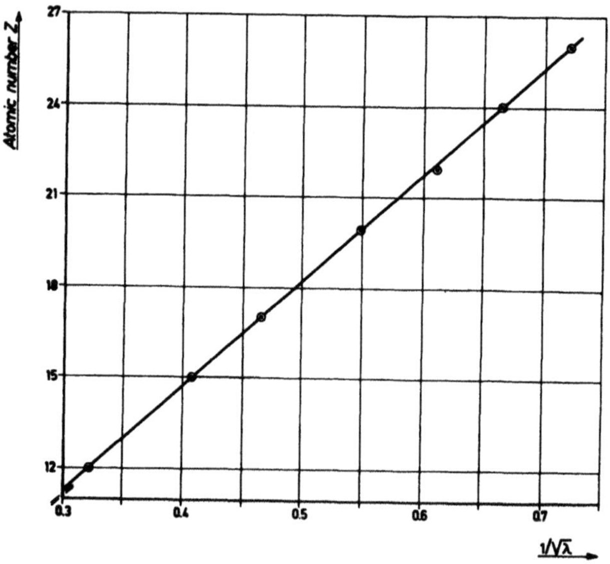

Fig.1

B The slope of the curve can be determined by taking two sets of ordinates, e.g.

$$X_1/Y_1 = 0.3/11.35$$

$$X_2/Y_2 = 0.725/26.18$$

From which the slope is

$$\frac{Y_2 - Y_1}{X_2 - X_1} = \frac{14.83}{0.425} = 34.96$$

As K is 1/slope, K = 0.02860 or 2.87 x 10^{-2}

From the curve shown σ is equal to 0.88.

The wavelength for manganese Kα (Z = 25) is thus

$$\lambda(\mathrm{Mn\,K}\alpha) = \frac{1}{K^2 \cdot (Z - \sigma)^2} = 2.101\ \text{Å} \quad (\text{true value} = 2.103\ \text{Å})$$

QUESTION 2

Characteristic spectra - Prediction of possible transitions

The electronic structure of Cu ($Z = 29$) is denoted $1s^2$, $2s^22p^6$, $3s^23p^63d^9$, $4s^2$.
The number preceeding each quantum group, i.e. 1, 2 or 3 is the group quantum number n. s, p and d represent the angular quantum numbers, having values of 0, 1 or 2 respectively. The small index number is the number of electrons in the indicated orbitals. Thus $3p^6$ represents $n = 3$ (i.e. M orbital), p = 1 (ℓ quantum number) and 6 electrons in the p orbitals. The spin quantum number is denoted s.
The vector sum of $\overline{\ell} + \overline{s}$ is called $\overline{J}$. The value of J must always be positive.

Given the selection rules $\Delta\ell = \pm 1$ and $\Delta J = 0$ or ± 1, list the possible transitions of copper, in its ground state, for the L series.

The transition groups L_I, L_{II} etc. are made by combining the ℓ and J quantum numbers as in the table given. Indicate the change in group for each of the transitions you propose and use the tables at the end of the book to give each transition its accepted nomenclature.

Transition group	ℓ	J	Transition group	ℓ	J
L_I	0	1/2	M_{III}	1	3/2
L_{II}	1	1/2	M_{IV}	2	3/2
L_{III}	1	3/2	M_V	2	5/2
M_I	0	1/2	N_I	0	1/2
M_{II}	1	1/2			

ANSWER 2

The given electronic structure of copper can be written as follows. Since it is given that the first selection rule is $\ell = \pm 1$ only certain transitions are allowed:

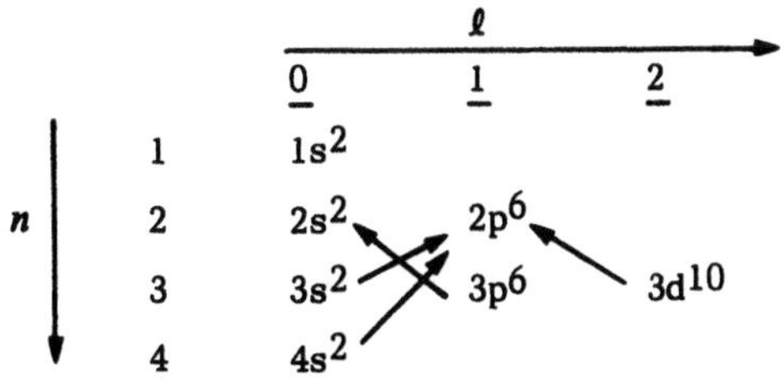

By listing these transitions with possible values of ℓ, s and J before and after transition and then applying the selection rules all possible transitions can be identified.

TRANSITION (only $\Delta\ell = 1$)		INITIAL STATE ℓ	s	J	group	FINAL STATE ℓ	s	J	group	ΔJ	LINE
3p →	2s	1	$+\frac{1}{2}$	3/2	M_{III}	0	$+\frac{1}{2}$	$+\frac{1}{2}$	L_I	-1	$\beta 3$
3p →	2s	1	$+\frac{1}{2}$	3/2		0	$-\frac{1}{2}$	$-\frac{1}{2}$			negative J
3p →	2s	1	$-\frac{1}{2}$	$\frac{1}{2}$	M_{II}	0	$+\frac{1}{2}$	$+\frac{1}{2}$	L_I	0	$\beta 4$
3p →	2s	1	$-\frac{1}{2}$	$\frac{1}{2}$		0	$-\frac{1}{2}$	$-\frac{1}{2}$			negative J
3d →	2p	2	$+\frac{1}{2}$	3/2	M_V	1	$+\frac{1}{2}$	3/2	L_{III}	-1	$\alpha 1$
3d →	2p	2	$+\frac{1}{2}$	3/2		1	$-\frac{1}{2}$	$\frac{1}{2}$		2	J = 2
3d →	2p	2	$-\frac{1}{2}$	3/2	M_{IV}	1	$+\frac{1}{2}$	3/2	L_{III}	0	$\alpha 2$
3d →	2p	2	$-\frac{1}{2}$	3/2	M_{IV}	1	$-\frac{1}{2}$	$\frac{1}{2}$	L_{II}	1	$\beta 1$
3s →	2p	0	$+\frac{1}{2}$	$\frac{1}{2}$	M_I	1	$+\frac{1}{2}$	3/2	L_{III}	+1	ℓ
3s →	2p	0	$+\frac{1}{2}$	$\frac{1}{2}$	M_I	1	$-\frac{1}{2}$	$\frac{1}{2}$	L_{II}	0	n
3s →	2p	0	$-\frac{1}{2}$	$-\frac{1}{2}$		1	$+\frac{1}{2}$	3/2			negative J
3s →	2p	0	$-\frac{1}{2}$	$-\frac{1}{2}$		1	$-\frac{1}{2}$	$\frac{1}{2}$			negative J
4s →	2p	0	$+\frac{1}{2}$	$\frac{1}{2}$	N_I	1	$+\frac{1}{2}$	3/2	L_{III}	+1	$\beta 6$
4s →	2p	0	$+\frac{1}{2}$	$\frac{1}{2}$	N_I	1	$-\frac{1}{2}$	$\frac{1}{2}$	L_{II}	0	$\gamma 5$
4s →	2p	0	$-\frac{1}{2}$	$-\frac{1}{2}$		1	$+\frac{1}{2}$	3/2			negative J
4s →	2p	0	$-\frac{1}{2}$	$-\frac{1}{2}$		1	$-\frac{1}{2}$	$\frac{1}{2}$			negative J

QUESTION 3

Continuous spectra - use of Kramers' formula

Plot the approximate distribution of intensity that would be obtained from a chromium anode X-ray tube operated at 50 kV. Use Kramers' formula[4] to give the relative intensity in the number of photons (I) over the wavelength range 0-5 Å.

$$I(\lambda)\cdot d\lambda = K\cdot i\cdot Z\cdot\left(\frac{\lambda}{\lambda_0}-1\right)\frac{1}{\lambda_2}\cdot d\lambda \qquad (2)$$

where $I(\lambda)$ is the intensity at wavelength λ and λ_0 is the minimum wavelength of the continuum. Z, the atomic number of the anode, and i, the tube current, are both constant values and can be joined with K (a constant) to give an overall constant K'.

In practice the tube spectrum is considerably modified due to self absorption by the anode itself. Correct the tube spectrum plotted, with the following values for the fraction of radiation (E_e) emitted by the anode at a given take-off angle of the tube. (These values were obtained using the heat dissipation method of Spencer[5])

Wavelength (Å)	(E_e)	Wavelength (Å)	(E_e)
0.375	1.00	2.25	0.75
0.50	0.96	2.50	0.69
1.00	0.70	3.00	0.63
1.50	0.51	4.00	0.46
2.00	0.30	5.00	0.29

The K absorption edge of chromium is at 2.07 Å.

Check that the maximum of the continuous comes at the right value by differentiating Kramers' formula to obtain the maximum.

ANSWER 3

In order to obtain the required spectra values of $I(\lambda)$ must be calculated for different wavelengths between 0 - 5 Å using Kramers' formula

$$I(\lambda) \propto \left(\frac{\lambda}{\lambda_0} - 1\right)\frac{1}{\lambda^2}$$

The values of $I(\lambda)$ can then be multiplied by the given values of (E_e) to give the spectrum corrected for self absorption.

The minimum wavelength can be calculated using the relationship:

$$\lambda_0 = \frac{12.4}{V}$$ where V is the kilovoltage on the X-ray tube.

In this case: $\frac{12.4}{50} = 0.248$ Å

λ	λ/λ_0	$(\lambda/\lambda_0 - 1)$	λ^2	$1/\lambda^2$	$I(\lambda)$	$E_e \times I(\lambda)$
0.375	1.515	0.515	0.141	7.09	3.56	3.56
0.50	2.01	1.01	0.25	4	4.04	3.88
1.00	4.03	3.03	1	1	3.03	2.12
1.50	6.05	5.05	2.25	0.445	2.25	1.14
2.00	8.06	7.06	4	0.25	1.762	0.528
2.25	9.07	8.07	5.06	0.198	1.60	1.20
2.50	10.07	9.07	6.25	0.16	1.45	1.00
3.00	12.09	11.09	9	0.111	1.232	0.776
4.00	16.12	15.12	16	0.0625	0.945	0.435
5.00	20.15	19.15	25	0.04	0.766	0.222

The curves given in Fig.2 were obtained using the values given in the above table.

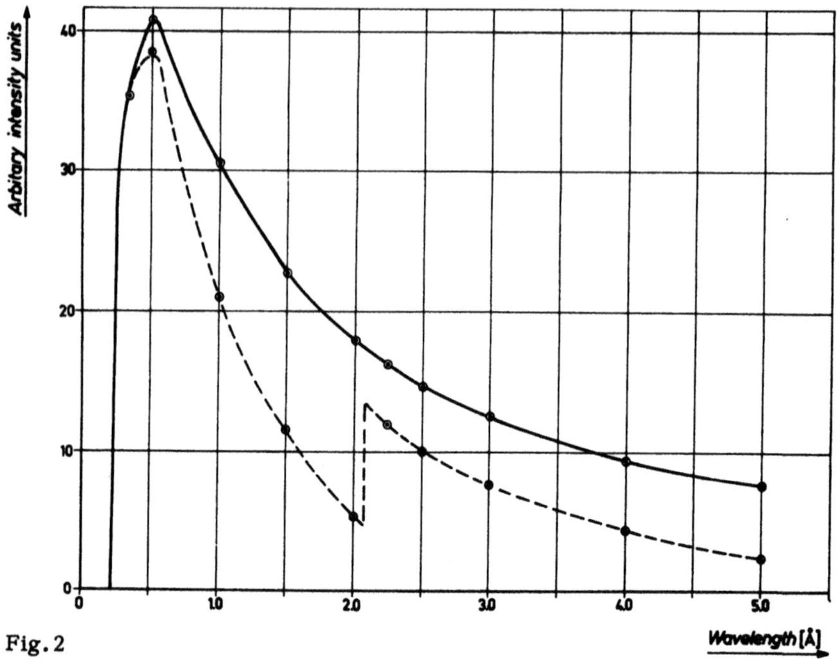

Fig. 2

The maximum of the curve can be obtained by differentiating Kramers' formula to obtain dI/dλ and equating this to zero, i.e.,

$$I(\lambda)=\left(\frac{K}{\lambda\lambda_0}-\frac{K}{\lambda^2}\right)$$

$$\frac{dI}{d\lambda}=\left(-\frac{1}{\lambda^2\lambda_0}+\frac{2}{\lambda^3}\right)=0 \quad \text{for turning point}$$

i.e.,

$$\frac{2}{\lambda^3}=\frac{1}{\lambda^2\lambda_0} \quad \text{or } \lambda_{max}=2\lambda_0$$

i.e. λmax = about 0.50 A

QUESTION 4

Continuous radiation - Scattered radiation

A The continuous radiation from an X-ray tube is scattered by the sample, diffracted by the analyzing crystal and measured by the detector.

Calculate the angle at which the continous radiation starts for a tungsten anode X-ray tube and an LiF (200) analyzing crystal, for which 2d = 4.028 Å. The X-ray tube is operated at 100 kV, with a constant potential generator.

B An attempt is made to measure the true value of the kilovoltage on the X-ray tube (meter reading 100 kV) by using a low atomic number sample to obtain a high intensity scattered continuum which is then extrapolated to zero intensity. Using the same method as in the first part of the question a value of about 83.6 kV is obtained. An independent check shows the 100 kV meter reading to be correct within 0.2%. Can you explain the reason for the large error in the estimated value of 83.6 kV.

ANSWER 4

A It is stated that the X-ray tube is operated at 100 kV, so the equivalent minimum wavelength of the continuous radiation will be:

$$\lambda_{min} = \frac{12.4}{100} = 0.124\,\mathring{A}$$

Substitution of this value in the Bragg law will give the equivalent angle:

$$n\lambda = 2d\sin\theta \quad \text{or} \quad \theta = \sin^{-1}\left(\frac{n\lambda}{2d}\right)$$

$$= \sin^{-1}\left(\frac{1 \times 0{\cdot}124}{4{\cdot}028}\right) \text{ or } \theta = 1.75° \text{ and } 2\theta = 3.50°$$

B The large difference between the theoretical and measured values of tube kilovoltage is simply an effect of the Compton shift[6]. Due to the use of a low atomic number sample much of the scatter is incoherent. A low atomic number element contains many electrons that are only loosely bound, and when an impinging X-ray photon strikes one of these electrons part of the energy of the photon is transferred to the electron. Since the total momentum remains unchanged the relationship between the wavelength of the impinging X-ray photon (λ_o) and the same photon after being incoherently scattered (λ_c) can be shown to be:

$$\lambda_c - \lambda_o = 0.0243(1 - \cos\psi)$$

where ψ is the scattering angle (normally about 90° in a conventional X-ray spectrometer) thus the Compton shift is 0.0243 Å. Fig.3 shows a plot of intensity against wavelength for a tungsten anode X-ray tube operated at 100 kV and illustrates the Compton shift effect. (The angular shift is also indicated for a LiF (200) crystal).

Thus in the given case where a low atomic number sample is used, instead of obtaining the mimum of the continuum at 3.50° 2θ it actually

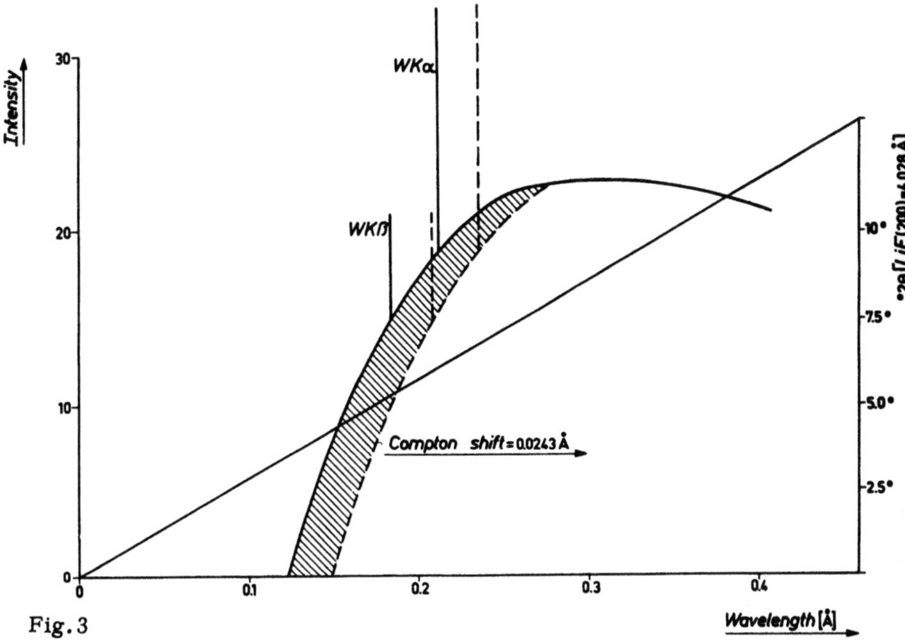

Fig. 3

occurs about 4.20° 2θ i.e. at a wavelength of 0.1483 Å given by

$$\underset{100\ \text{kV}}{0.124} + \underset{\text{Compton}}{0.0243} = 0.1483\ \text{Å}$$

The apparent kilovoltage equivalent to this wavelength is:

$$\frac{12.4}{0.1483} = 83.6\ \text{kV}$$

QUESTION 5

Detector resolution

The resolution R of a proportional counter is expressed in terms of the peak width at half height multiplied by 100, and divided by the value of the maximum pulse amplitude.

The theoretical value[7] of R for a given energy E is:

$$R = \frac{38.3 \times K}{\sqrt{E}} \tag{3}$$

for an argon filled gas proportional counter.

Here K is a factor taking a value between 1 and 1.5 depending upon the cleanliness of the counter[7].

Measurements on the Fe Kα wavelength (E = 6.41 keV) gave a peak width at half height of 115 scale divisions and a pulse amplitude of 580 scale divisions.

A What is the measured resolution of the counter for Fe Kα radiation

B What is the value of K in this instance

C What would be the measured pulse amplitude of Al Kα (E = 1.49 keV) in scale divisions, using the same conditions

D What would be the measured resolution for Al Kα

ANSWER 5

A The resolution of the counter is:

$$R = \frac{\text{peak width at half height x 100}}{\text{maximum pulse amplitude}} = \frac{115 \times 100}{580} = \underline{19.8\%}$$

B Since it is given that

$$R = \frac{38.3 \times K}{\sqrt{E}}$$

subsitution of the values for R (19.8) and E (6.41) gives:

$$K = \frac{R\sqrt{E}}{3.83} = \frac{19.8\sqrt{6.41}}{38.3} = 1.31$$

C Since a proportional counter is being used the ratio of the pulse amplitudes for two given energies will be directly proportional to the ratio of the energies. Thus:

$$\text{Pulse amplitude Al K}\alpha = \text{pulse amplitude Fe K}\alpha \times \frac{\text{energy Al K}\alpha}{\text{energy Fe K}\alpha}$$

$$= 580 \times \frac{1.49}{6.41} = \underline{\text{135 scale divisions}}$$

D It will be apparent from the given equation that R is inversely proportional to the square root of the energy for a given counter with a given K value. Thus:

$$R_{Al\,K\alpha} = R_{Fe\,K\alpha}\sqrt{\frac{E_{Fe\,K\alpha}}{E_{Al\,K\alpha}}}$$

which gives

$$19.8\sqrt{\frac{6.41}{1.49}} = \underline{41\%}$$

QUESTION 6

Dispersion as a function of 'd' spacing

LiF is commonly used as an analysing crystal. The (200) reflection planes are commonly used, having a "$2d$" value of 4.028 Å. It is, however, possible to use reflecting planes with other Miller indices (hkl) for example the (220), the (420) and the (422) planes.

LiF crystalizes with (face-centred) cubic symmetry for which the relation between the reticular distances, d_{hkl}, the Miller indices (hkl) and the unit cell dimension a_0 is

$$d_{hkl} = \frac{a_0}{\sqrt{h^2+k^2+l^2}} \qquad (4)$$

A Calculate the value of a_0 and the $2d$ value of the (220), (420) and (422) planes.

B The dispersion of a crystal $\frac{d\theta}{d\lambda} = \frac{n}{2d}\frac{1}{\cos\theta}$ (5)

Sn Kα and Sb Kα radiations have wavelengths of 0.492 Å and 0.472 Å, respectively. Calculate the angular dispersion of these two wavelengths with the four different cuts of LiF (i.e. (200), (220), (420) and (422)). Remember that the goniometer position is in terms of the angle 2θ.

ANSWER 6

A The following relationship is given

$$d_{hkl} = \frac{a_0}{\sqrt{h^2+k^2+l^2}}$$

The $2d$ value of the (200) is 4.028 Å, so

$$2d = 4.028 \therefore d = 2.014 = \frac{a_0}{\sqrt{4+0+0}} = \frac{a_0}{2}$$

$$\therefore a_0 = 4.028 \text{ Å}$$

The $2d$ values for the other cuts will be:

$$(220)\quad 2d = \frac{2 \times 4.028}{\sqrt{4+4+0}} = \frac{8.056}{\sqrt{8}} = 2.850 \text{ Å}$$

$$(420)\quad 2d = \frac{2 \times 4.028}{\sqrt{16+4+0}} = \frac{8.056}{\sqrt{20}} = 1.802 \text{ Å}$$

$$(422)\quad 2d = \frac{2 \times 4.028}{\sqrt{16+4+4}} = \frac{8.056}{\sqrt{24}} = 1.645 \text{ Å}$$

B It is stated that the dispersion $\frac{d\theta}{d\lambda} = \frac{n}{2d}\frac{1}{\cos\theta}$

Thus the angular dispersion for a certain value of dλ will be:

$$d\theta = \frac{n\, d\lambda\, 57.3 \sec\theta}{2d}$$

The 57.3 is necessary for the conversion to radian measure. Since in this instance dλ = (0.492 − 0.472) = 0.02 Å and n = 1,

$$\therefore d\theta = \frac{0.02 \times 57.3 \times \sec\theta}{2d}$$

The value of θ may be calculated using the Bragg law $n\lambda = 2d\sin\theta$.

crystal cut	2d(Å)	θ° (Sn Kα)	sec θ	$d\theta^{\circ} = \frac{1.146 \sec\theta}{2d}$	d (2θ)°
(200)	4.028	7.0	1.0075	0.287	0.57
(220)	2.850	9.94	1.0152	0.408	0.82
(420)	1.802	15.84	1.0395	0.663	1.32
(422)	1.645	17.40	1.0480	0.732	1.46

QUESTION 7

Choice of collimator/crystal combination

In the measurement of a mixture of elements a problem of the overlap of two lines is encountered. The lines can be separated by use of the following combinations:

(i) A topaz crystal (having good resolution but poor reflecting power) plus a coarse collimator, placed between sample and crystal.

(ii) A LiF (200) crystal (having poor resolution but good reflectivity) plus a fine collimator, placed between sample and crystal.

The topaz crystal gave the following intensities:

	peak (c/s)	background (c/s)
fine collimator	487	68
coarse collimator	1301	285

Using the same instrumental conditions a LiF (200) crystal was found to increase the peak and background count rates by a factor of 1.41.

Which of the crystal/collimator combinations is the best

ANSWER 7

Although as shown in the following table there are four possible combinations of collimator and crystal, only two of these are under consideration, namely (b) and (c). Inspection of Fig.4 reveals why this is so.

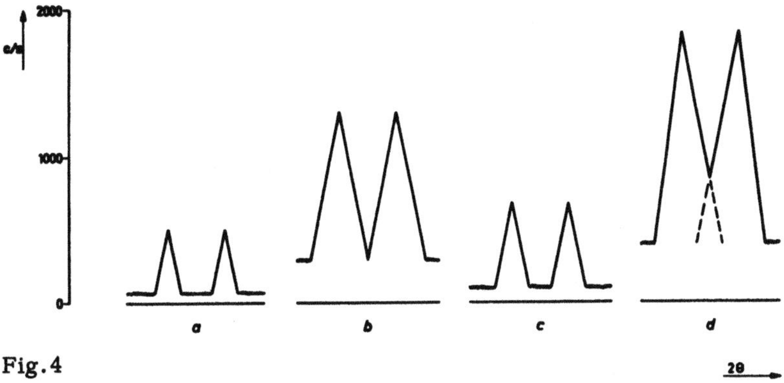

Fig.4

Combination (a) gives too much resolution at the expense of too great a loss of intensity, whereas combination (d) shows very high intensity but insufficient resolution.

Choices:

(a)	fine	topaz	good resolution/poor reflection
(b)	coarse	topaz	
(c)	fine	LiF (200)	poor resolution/good reflection
(d)	coarse	LiF (200)	

The percentage error (ε%) incurred in counting where peak (R_p) and background (R_b) count rates are involved is given by:

$$\varepsilon\% = \frac{100}{\sqrt{T}} \cdot \frac{1}{\sqrt{R_p} - \sqrt{R_b}} \tag{6}$$

where T is the total analysis time. Thus for a fixed analysis time, ε is at a minimum when $(\sqrt{R_p} - \sqrt{R_b})$ is maximum. This expression is therefore taken as the figure of merit (F.O.M.). It is stated that the use of the LiF (200) crystal increases the values of R_p and R_b by 1.41, so the calculation of the figure of merit is as follows:

	R_p	$\sqrt{R_p}$	R_b	$\sqrt{R_b}$	F.O.M. = $\sqrt{R_p} - \sqrt{R_b}$
(b)	1301	36.1	285	16.9	19.2
(c)	487 x 1.41 = 687	26.2	68 x 1.41 = 96	9.8	16.4

Since (b) gives a F.O.M. greater than that of (c), (b) represents the optimum combination i.e. coarse collimator and topaz crystal.

QUESTION 8

Pulse height selection - crystal fluorescence

It is necessary to set up an X-ray spectrometer for the measurement of sodium in a series of rock specimens. An argon/methane flow counter is being employed and a KAP (potassium hydrogen phthalate) analysing crystal. The rock samples contain quite a lot of calcium and this leads to strong crystal fluorescence.

How many maxima will be seen in the pulse amplitude distribution from the flow counter. Plot out roughly the pulse amplitude distribution obtained* at the setting of the Na Kα line, where the peak to background ratio is about 1.6, assuming a detector resolution of 38% for Na Kα. Bear in mind that a potassium K escape peak wil' occur with an intensity roughly 10% of that of the photopeak.

Use the given detector resolution for Na Kα to calculate the resolution (R) for other energies (E) which occur, using the expression:

$$R = \frac{k}{\sqrt{E}} \tag{7}$$

where k is a constant.

* Since the detector resolution follows approximately a gaussian distribution it can in turn be approximately represented by a symmetrical triangle with a base equal to twice the peak width at half height. This triangle will then contain roughly 92% (2σ) of the intensity of the pulse distribution.

ANSWER 8

It is stated that crystal fluorescence will occur thus pulses will arise from both potassium K radiation and its associated escape peak. The situation in practice is likely to be slightly complicated by the presence of both K Kα and K Kβ radiation but compared to the former the K Kβ will be insignificant.

In order to plot the approximate pulse amplitude distribution the position of each pulse amplitude maximum and the width of each distribution at half height must be known.

The peak position will occur at values directly proportional to the energies (*E*) of the radiations, and so an energy scale can be used directly. The K escape pulses will occur with a maximum equivalent to the difference in the energy of K Kα radiation and the Ar Kα radiation which has escaped (2.96 keV).

The width of the distribution at half height (*W*) can be calculated from the detector resolution *R*, as follows

$$R = \frac{W}{E} \times 100\,\% \quad \text{or } W = \frac{R \times E}{100}$$

The detector resolution for Na Kα is given as 38% and the following equation can be used to calculate resolutions for K Kα and the K escape peak:

$$R = \frac{k}{\sqrt{E}} \quad \text{or for Na K}\alpha\ 38\% = \frac{k}{\sqrt{1.04}} \quad \therefore\ k = 38.7$$

$$\text{thus } R(\text{K K}\alpha) = \frac{38.7}{\sqrt{3.309}} = 21.3\,\% \quad R(\text{K escape}) = \frac{38.7}{\sqrt{0.349}} = 65.7\,\%$$

it is also stated that the pulse amplitude distributions may be represented by triangles with a base width equal to twice the peak width and half height (*W*). Thus the (lower) level at which a distribution ends will be (*E*-*W*) and the (upper) level at which a distribution ends will equal (*E*+*W*). From the given peak to background ratio it is known that the K Kα will be approximately half as big as the Na Kα. Also stated is that the escape peak is roughly 10% as intense as the photopeak. The following data can now be used to sketch the pulse distribution:

	Na Kα	K Kα	K escape peak
Energy (keV)	1.040	3.309	0.349
Resolution (%)	37.6	21.3	65.7
$W = \frac{R \times E}{100}$	0.39	0.70	0.23
lower level (keV)	0.69	2.61	0.12
upper level (keV)	1.43	4.01	0.58

Figure 5 illustrates the constructed theoretical curve and shows also the actual curve obtained in practice.

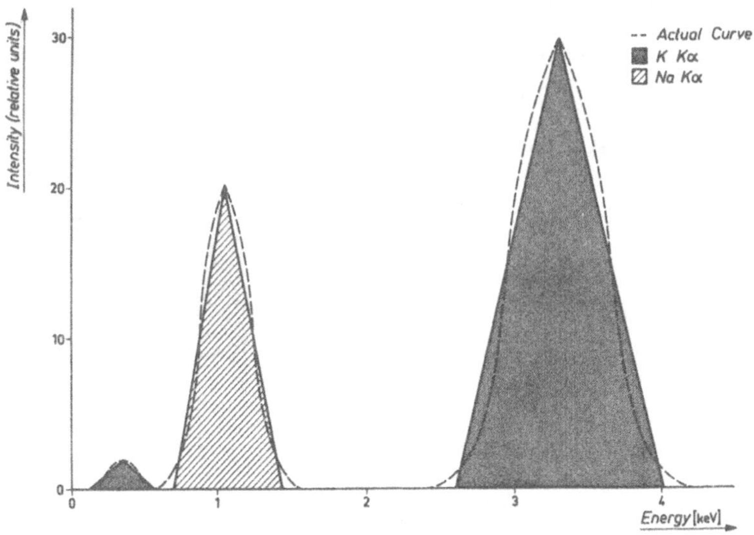

Fig. 5

QUESTION 9

Pulse height selection - pulse height variation

A mixture of sodium silicates and sodium phosphates is being analysed for phosphorus. The experimental conditions used are: tungsten anode X-ray tube, 50 kV-40 mA; coarse collimator; gypsum analysing crystal; gas flow proportional counter (Ar/10% methane) with pulse height selection.

On setting up the pulse height selector it is found that the maximum of the pulse amplitude distribution occurs at about 31 scale divisions for high phosphorus concentrations but at about 34 scale divisions for low phosphorus concentrations.

Lowering the gas gain of the counter (to remove the effect of count rate-pulse shift dependance) does not eliminate the problem.

Explain the reason for the effect and offer a possible solution.

ANSWER 9

The most usual explanation for variation of pulse amplitude with concentration (i.e. implied variation of count rate) is that due to lowering of the field around the anode due to a large number of positive ions. However, evidently this is not the case here.

The effect is in fact due simply to crystal fluorescence. A gypsum ($CaSO_4$) crystal is being employed and the large amount of radiation scattered by the low average atomic number matrix causes excitation of a significant amount of characteristic sulphur radiation. Thus when the spectrometer is set at the angle for phosphorus Kα two distributions occur in the pulse height diagram (see Fig.6) but the counter is not good enough to resolve them.

Thus at low phosphorus concentrations the sulphur radiation is the dominant feature of the unresolved pulse amplitude, which peaks at a point just lower than the voltage corresponding to S Kα, i.e. at about 34 scale divisions.

At high phosphorus concentrations, however, phosphorus Kα becomes the dominant feature with subsequent shift towards the voltage corresponding to P Kα, and the unresolved distribution now peaks at about 31 scale divisions.

The solution to this problem is simply to avoid use of the gypsum crystal. Either penta-crythritol or ethylene-diamine d-tartrate would offer better alternatives.

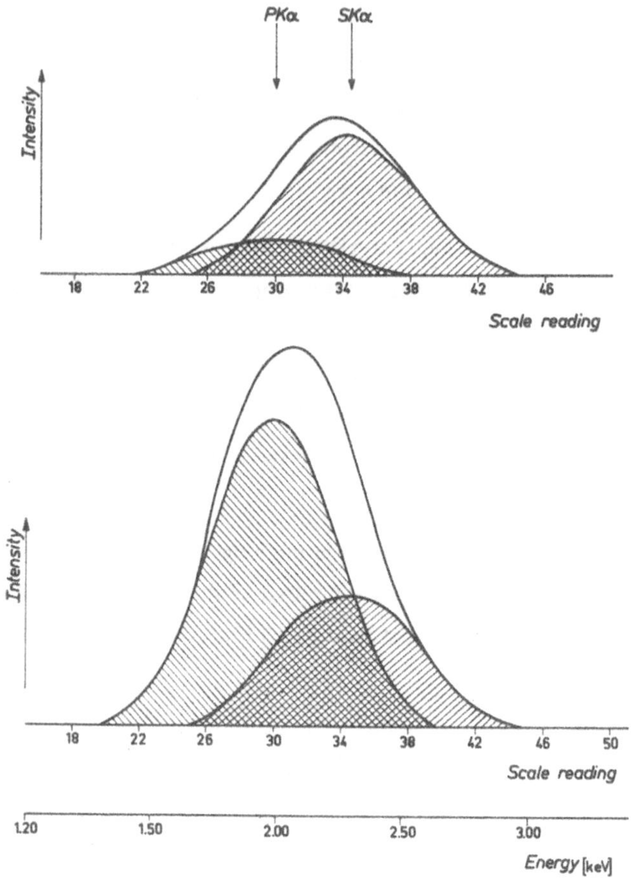
PKα
SKα
Intensity
18
22
26
30
34
38
42
46
Scale reading
Intensity
18
22
26
30
34
38
42
46
50
Scale reading
1.20
1.50
2.00
2.50
3.00
Energy [keV]

Fig. 6

QUESTION 10

Detectors - double plateaux

A series of measurements are made in which counting rate is plotted as a function of counter voltage. The counter in this case is an argon/methane gas flow proportional counter. A single plateau is found for a sample of pure sulphur, using the S Kα line, but pure samples of iron and chromium are both found to give a second plateau, again when measuring their Kα lines.

Give a possible explanation for the double plateau. Why does not sulphur also show this effect.

ANSWER 10

The effect described is essentially an escape peak phenomena. When the radiation entering the counter is energetic enough to eject an argon K electron (i.e. about 3.22 keV) argon Kα radiation is produced, which escapes from the counter. Thus instead of a single pulse amplitude distribution being formed, with a maximum V proportional to the energy E of the incident radiation, two pulse amplitudes are formed, one with a maximum at V and one with a maximum at V', where V' is proportional to $(E - E_{\text{ArK}\alpha})$. This second pulse amplitude is called the escape peak. As shown in the upper part of Fig.7 the escape peak (at about 100 mV) is roughly 10% the intensity of the photo-peak (400 mV). The counting rate is proportional to the area under the distribution.

As the counter voltage is increased all pulses are enlarged, but at low voltage levels are insufficiently large to pass the threshold level. This corresponds to about 50 mV on the upper part of Fig.7. Thus with a counter voltage of 1470 V, line A, the pulses are too small to pass the threshold. Increase of the voltage to 1525 V brings the photo pulses to a large enough size to pass the threshold and the count rate recorded corresponds to the area under the photo peak. The escape peak is, however, still too small to pass the threshold. This situation remains up to a voltage of about 1550 V, line C, and so there is a plateau between B and C.

Further increase in voltage brings the escape peak amplitude also over the threshold level with a further increase in count rate (the count rate scale on the ordinate of the lower curve is not linear) proportional to the area below the escape peak curve.

Further increase in voltage beyond about 1600 V, line D, gives no further increase in count rate since no further pulses are brought above the threshold and the count rate registered is always proportional to the number of pulses above the threshold level.

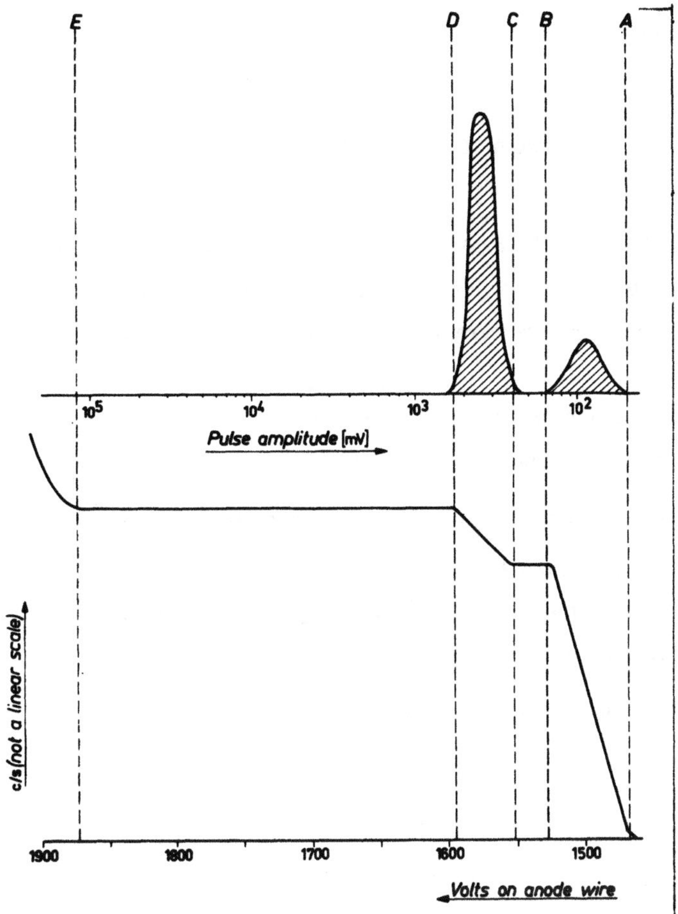

Fig.7

Increase of the voltage beyond line E takes the counter into the glow discharge region and the counter is no longer usuable.

Both chromium (5.42 keV) and iron (6.41 keV) have sufficient energy to excite argon Kα and thus show the effect, but sulphur (2.31 keV) is insufficiently energetic to excite argon Kα. In fact the effect is smaller in the case of iron since the absorption by argon of Fe Kα is less than that of Cr Kα, leading to a reduced photopeak.

QUESTION 11

Choice of detectors

A A choice has to be made between the scintillation counter and the flow proportional counter for the measurement of a certain wavelength. The following data were obtained under similar conditions:

	peak count rate	background count rate
scintillation counter	1600 c/s	40 c/s
proportional counter	400 c/s	4 c/s

which counter is best suited for this type of analysis.

B Using the data for the chosen detector, how many seconds should be used accumulating counts on the background, assuming a total analysis time of 148 seconds.

C What is the percentage error in the analysis (1σ).

ANSWER 11

A It can be shown that the statistical counting error (ε) is related to the peak (R_p) and background (R_b) counting rates in the following way:

$$\varepsilon\% = \frac{100}{\sqrt{T}} \cdot \frac{1}{\sqrt{R_p} - \sqrt{R_b}} \qquad (6)$$

where (T) the total analysis time is divided between peak counting time (T_p) and background counting time (T_b) as follows:

$$\frac{T_p}{T_b} = \sqrt{\frac{R_p}{R_b}} \qquad (8)$$

In the first equation ε is at a minimum when $\sqrt{R_p} - \sqrt{R_b}$ is maximum assuming, of course, that the analysis time is constant.

Thus for the scintillation counter $\sqrt{R_p} - \sqrt{R_b} = \sqrt{1,600} - \sqrt{40} = 33.7$ and for the proportional counter $\sqrt{400} - \sqrt{4} = 18$.

therefore the scintillation counter is the better option.

B Since $T = T_p + T_b$ substitution for T_b in terms of T_p and $\sqrt{R_p/R_b}$ will give an expression for Tp in terms of $\sqrt{R_p/R_b}$ and T.

From equation (8) $T_b = T_p \sqrt{R_b/R_p}$

giving $T = T_p + T_p \sqrt{R_b/R_p}$

or $$T_p = \frac{T}{1 + \sqrt{R_b/R_p}}$$

For the example given, $$T_p = \frac{148}{1 + \sqrt{1/40}} = \frac{148}{1.158} = 128 \text{ s}$$

$$T_b = 20 \text{ s}$$

C Finally $\varepsilon\% = \frac{100}{\sqrt{148}} \cdot \frac{1}{33.7} = 0.245\%$

QUESTION 12

Count rate differences with Geiger and proportional counters

A silicon specimen is being examined using nickel filtered Cu Kα radiation. The intensities of two reflections are measured with a Geiger-Müller counter and the following counting rates obtained:

	Peak A	Peak B
Geiger-Müller counter	1,400 c/s	200 c/s

However, when the measurements are checked with a xenon filled sealed proportional counter the count rates now obtained are:

	Peak A	Peak B
proportional counter	2,110 c/s	210 c/s

Which of the measurements is most likely to yield the correct line intensity ratio of A : B. Explain the difference between the Geiger-Müller counter and proportional counter results.

ANSWER 12

This is simply an effect of dead time and the Geiger-Müller counter results are the ones subject to error. Thus the proportional counter results are most likely to give the correct line intensity ratio.

The dead time of the Geiger-Müller counter is of the order of 240 μs compared with about 2 μs for the proportional counter. Dead time correction can be made using the expression:

$$R^{true} = \frac{R^{measured}}{1 - R^{measured} \cdot \tau} \qquad (9)$$

where (R^{true}) is the true count rate, ($R^{measured}$) the measured count rate and τ the dead time.

Thus for the Geiger-Müller counter count rate on peak A:

$$R^{true} = \frac{1\,400}{1 - (1\,400 \times 240 \times 10^{-6})} = 2\,110 \text{ c/s}$$

and for the proportional counter count rate on peak A:

$$R^{true} = \frac{2110}{1 - (2110 \times 2 \times 10^{-6})} = 2118 \text{ c/s}$$

Thus substantially the same count rate is obtained after correction, but the large effect of dead time in the case of the Geiger-Müller counter is clearly seen. Similar correction of the data for line B gives:

R^{true} (Geiger-Müller counter) line B = 210 cps

R^{true} (proportional counter) line B = 210 cps

QUESTION 13

Dispersion of the diffractometer

The dispersion of the diffractometer ($d\theta/d\lambda$) can be expressed in the given form obtained by differentiating the Bragg law $n\lambda = 2d\sin\theta$

$$\frac{d\theta}{d\lambda} = \frac{n}{2d} \cdot \frac{1}{\cos\theta} \quad (5)$$

where the angular separation ($d\theta$) of two wavelengths separated by ($d\lambda$) is expressed in terms of the diffraction angle θ for a certain spacing d. n is an integer which can be taken as equal to 1.

A Plot the angular separation of the Cu $K_{\alpha 1}$, Cu $K_{\alpha 2}$ doublet as a function of goniometer angle, over the range 0 to 100 $^{\circ}2\theta$.

The wavelengths of the Cu K lines are:

Cu $K_{\alpha 1}$ = 1.5405 Å

Cu $K_{\alpha 2}$ = 1.5443 Å

B Use the curve to establish the angular separation of the Cu $K_{\alpha 1}$, Cu $K_{\alpha 2}$ wavelengths by two of the d values of fluorite

(i) (220) plane d = 1.931 Å

(ii) (422) plane d = 1.115 Å

ANSWER 13

A The angular separation $d\theta$ must be calculated as a function of 2θ. Given that:

$$\frac{d\theta}{d\lambda} = \frac{n}{2d} \cdot \frac{1}{\cos\theta}$$

Substitution of $\frac{n}{2d}$ can be made by use of the Bragg law:

$$\frac{d\theta}{d\lambda} = \frac{\sin\theta}{\lambda} \cdot \frac{1}{\cos\theta}$$

$$\text{or} \quad d\theta = \frac{d\lambda}{\lambda} \cdot \tan\theta$$

Substituting the given wavelength values, $d\lambda = 0.0038$ and $\lambda = 1.5418$, gives:

$$d\theta = \frac{0.0038}{1.5418} \cdot 57.3 \cdot \tan\theta$$

the 57.3 being to convert from radians into degrees.

Therefore

$$d\theta = 0.1412 \cdot \tan\theta$$

Values of $d\theta$ and θ can now be calculated remembering that the diffractometer always measures 2θ, thus the total separation is $2 \times d\theta$.

$2\theta^\circ$	θ°	$\tan\theta$	$(d\theta)^\circ = 0.1412 \tan\theta$	separation $= 2x(d\theta)^\circ$
10	5	0.0875	0.0123	0.025
20	10	0.1763	0.0249	0.050
30	15	0.2679	0.0378	0.076
40	20	0.3640	0.0513	0.103
50	25	0.4663	0.0659	0.132
60	30	0.5774	0.0815	0.163
70	35	0.7002	0.0988	0.198
80	40	0.8391	0.1184	0.237
90	45	1.0000	0.1412	0.282
100	50	1.1918	0.1682	0.336

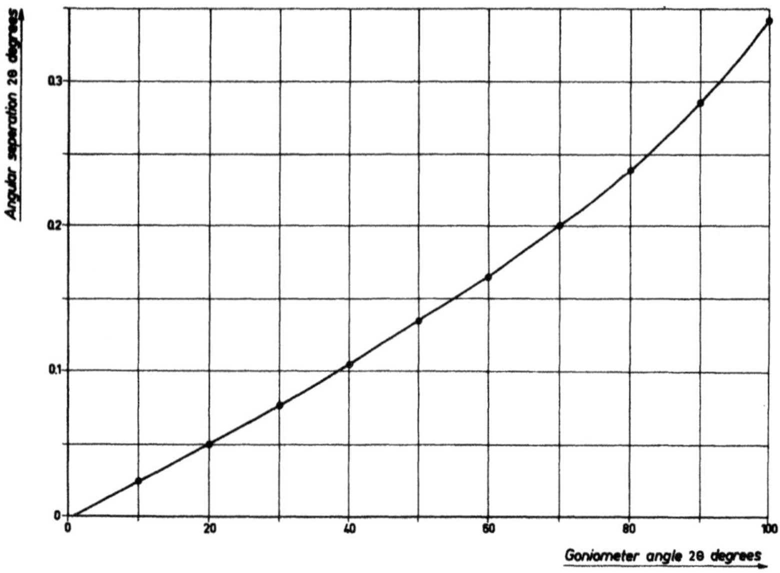

Fig. 8

The given curve was obtained by plotting the values in the first and last columns, (see Fig. 8)

B Two "d" values of fluorite are given and the angular separation of the $K\alpha_1$ and $K\alpha_2$ of Cu is to be calculated.

It is first necessary to calculate the angle at which the "d" values occur and then the appropriate values of the angular separation can simply be read off from the curve.

plane	d (Å)	$2d$ (Å)	$\theta^o = \sin^{-1}(\frac{\lambda}{2d})$	$2\theta^o$	separation ($^o2\theta$)
220	1.931	3.862	23.58	47.16	0.123
422	1.115	2.230	43.80	87.60	0.271

QUESTION 14.

Removal of sample fluorescence

The diffraction pattern of an iron specimen is being recorded with a copper target X-ray tube operating at 40 kV-50 mA. A Ni-filter is placed between the X-ray tube and specimen. A Xe proportional counter is used to record counts on one of the haematite reflections but even though this is one of the strongest reflections usable, the peak to background is only about 1.5 to 1.

The pulse height selector is to be used in order to remove iron fluorescence, but due to the close proximity of the Fe $K\alpha$ and Cu $K\alpha$ lines great care is required in setting up the pulse height selector. The measured resolution of the detector for Cu $K\alpha$ radiation is 13%.

A Plot the pulse amplitude distribution diagram using the abscissa scale in terms of keV. Any escape peaks that might arise following the excitation are roughly 10% of the intensity of the photopeaks.

B Assuming that an attenuator setting has been employed giving a pulse range of 5 to 10 keV, within the 0 to 100 unit scale of the pulse height selector, select approximate lower level and window settings that will give the optimum peak to background ratio.

C Another solution is to use a Cr anode X-ray tube operated at 40 kV 30 mA, again with the pulse height selector. Would this be expected to give better results than those obtained in B. What more could be done to use the relatively long wavelength Cr $K\alpha$ radiation with more efficiency.

ANSWER 14

The majority of the background arises from Fe K radiation excited within the sample by the incident Cu Kα radiation. Thus the pulse diagram will include distributions from Cu Kα, Fe Kα and Fe Kβ. Since the peak to background is stated to be about 1.5 to 1 the distribution for Cu Kα will be roughly half as big again as for Fe Kα.

A The pulse diagram can be constructed with knowledge of the resolution of the detector for each of the radiations measured. The resolution (***R***) of the detector is given by:

$$R(\%) = \frac{K}{\sqrt{E}} \qquad (7)$$

where E is the energy of the measured radiation. We are told that R is 13% for Cu Kα (E = 8.04 keV) hence K must equal 37. The resolution R can also be expressed in terms of the peak width W at half height for an energy peaking at V, i.e.,

$$R = \frac{W}{V} \cdot 100\ \%$$

Thus knowing the value of R and the value of V (i.e. E), W can be estimated. Thus:

	Cu Kα	Fe Kα	Fe Kβ
λ (Å)	1.542	1.937	1.757
E (keV)	8.04	6.41	7.07
R (%)	13	14.6	13.9
W (keV)	1.045	0.935	0.983
lower level (keV)	7.00	5.48	6.09
upper level (keV)	9.09	7.35	8.05

The base width of the distribution will equal approximately 2 W, therefore the distribution will begin (lower level) at $(E-W)$ and end (upper level) at $(E+W)$.

The given figure 9 was constructed using the data above.

Since Fe Kα and Cu Kα are both energetic enough to excite Xe Lα radiation (excitation potential 4.78 keV) the pulse diagram will also contain Fe Kα and Cu Kα escape peaks. The energies of these escape peaks will be simply the differences between the energy of Xe Lα and Cu Kα and Fe Kα respectively. Thus:

Fe Kα (escape peak) = 6.41 − 4.14 = 2.27 keV

Cu Kα (escape peak) = 8.04 − 4.14 = 3.90 keV

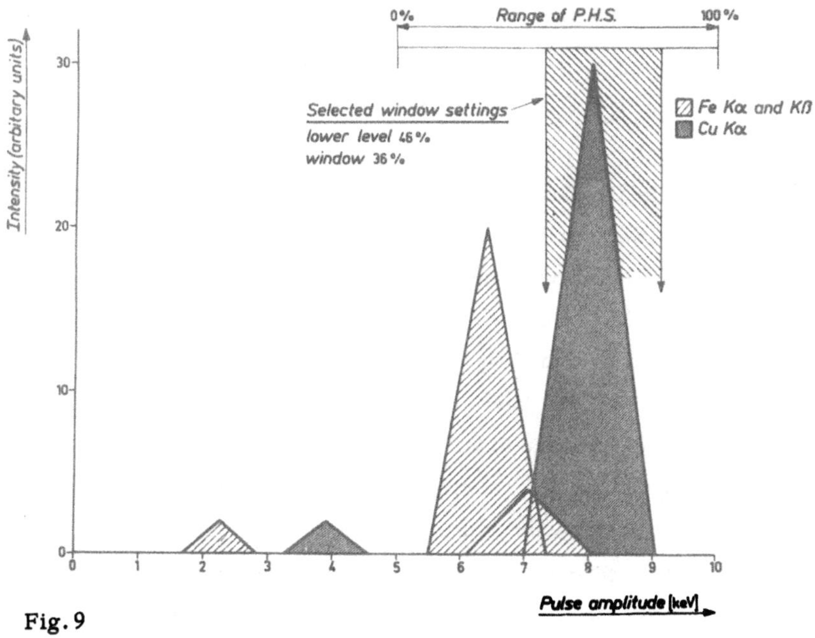

Fig. 9

B Since the range of the pulse height selector is 5 keV (0) to 10 keV (100) it is relatively simple to select lower level and window settings from the constructed figure.

The lower level must, however, be selected very carefully since too low a value will allow Fe Kα radiation to pass and too high a value will prevent passage of all of the Cu Kα. The presence of the Fe Kβ further complicates the issue but since this peak is almost at the exact mid-point of the Cu Kα and Fe Kα radiation distributions little can be done to remove its effect. From the given Fig.9 it will be seen that a setting of about 7.3 keV would be optimal, this in turn corresponding to 46%. The setting of the upper level is not at all critical and a value of 9.1 keV would be sufficient to embrace the whole Cu Kα distribution. Thus the window width would be (9.1 − 7.3) keV or 1.8 keV, corresponding to 36%.

Thus the optimum settings are:

lower level	46%
window	36%

Use of the Cr anode X-ray tube will certainly be expected to lead to the production of far less background radiation from specimens containing Fe. This is because whereas Cu Kα (λ = 1.542 Å) is able to excite Fe K radiation (Fe absorption edge 1.744 Å), the wavelength of Cr Kα (λ = 2.291 Å) is too long to cause excitation.

As far as the peak intensity is concerned more than 50% of the Cu Kα radiation is lost due to the fluorescence process, but against this 2 kW are used on the Cu anode tube but only 1.2 kW on the Cr anode tube. Also the absorption due to air is around 50% in the case of Cr Kα radiation as opposed to about 10% in the case of Cu Kα. When all of these effects are taken into consideration the use of the Cr anode tube gives marginally better results for the examination of specimens very high in Fe.

However this marginal improvement can be increased very considerably by use of a vacuum attachment on the goniometer to reduce the total absorption due to air to less than 10%. Thus something like a factor of 1.8 gain in the figure of merit can be obtained compared with similar measurements in air. Also the accuracy of the measurements can be improved considerably by use of a vacuum path for Cr Kα radiation, since changes in atmospheric pressure can lead to counting errors of up to 10%.

QUESTION 15

Spurious peaks in X-ray diffractograms

A diffractogram of quartz exhibited an additional spurious peak at approximately 6° (2θ). The effect was observed both in specimens containing only low atomic number elements and specimens containing high atomic number elements. The patterns were always taken using Cu radiation, proportional counter (Xe-filled) and pulse height selection. The pulse height selector was set at the 45% acceptance level since the resolution of the counter (peak width at half height) was 18% for Cu Kα radiation.

Explain the occurence of this peak for a pure quartz specimen, using the following experimental findings.

The (2θ) angle depends on the composition of the sample, the peak intensity increases with increasing high tension on the X-ray tube; below 34.5 kV the peak disappears completely; the angular position of the peak depends on the pulse height selector setting; using strictly monochromatic radiation, e.g. with a crystal monochromator, no peak in the region of 6°(2θ) is observed.

The three strongest lines for α-quartz are:

d (Å)	I/I_0
3.34	100
4.26	35
1.82	17

ANSWER 15

From the given experimental facts, it can be concluded that the pulse height selector and the reflection of the white spectrum from the X-ray tube by the crystal planes in the specimen, play a dominant part in the appearance of the extra peak. The disappearance below 34.5 kV strongly suggests a connection with the absorption of the X-rays in the proportional counter since the K-absorption edge of Xe occurs at 34.5 keV.

The reflection of a line from the white continuum should give pulses in the detector which cannot pass the pulse height selector, as their energy difference with the Cu Kα radiation is too great. However, at incident energies greater than 34.5 keV, an escape peak will occur.

The window width utilized was 45% of 8.04 keV = 3.6 keV, so that all energies between 6.2 and 9.8 keV can pass the window.

As far as the production of escape peaks is concerned, a line from the white continuum having an energy of 40 keV gives rise to an escape peak at

$$40 - 29.6\ (\text{Xe K}\alpha) = 10.4\ \text{keV}$$

The width at half height of this peak would be

$$18 \times \frac{8.04}{10.4} = 16\%.$$

The escape peak thus lies between 8.7 and 12.1 keV, assuming a symmetrical triangular distribution. The part below 9.8 keV can thus pass the window of the pulse height selector. An analogous calculation shows that for energies between approximately 35 and 45 keV a certain part of the escape peak can be registered. The maximum of this peak will correspond to that energy where the total escape peak falls between the window limits or coincides with Cu Kα, i.e. at

$$29.6 + 8 = 37.6\ \text{keV}$$

As the escape peak starts at energy above 34.5 keV, an asymmetric peak will result.

The 2θ values of this peak can be found as follows:

energy (keV)	λ (Å) = $\frac{12.4}{\text{energy}}$	reflection plane d (Å)	$\theta^{o} = \sin^{-1} \frac{\lambda}{2d}$	$2\theta^{o}$
		4.26	2.27	4.54
37.6	0.33	3.34	2.83	5.66
		1.82	5.20	10.40

QUESTION 16

Extra lines in X-ray diffractometry

A diffractometer trace is recorded of a compound known to contain a high concentration of Ba. The pattern is characterized by a very strong line with a *d*-spacing of 3.08 Å, plus many others. A weak, broad line was observed above a rather high background at approximately 6.5 $^{\circ}(2\theta)$. It was suspected from the shape of this line that it did not belong to the regular diffraction pattern. Can there be an alternative explanation for this line.

Experimental conditions:

Diffractometer, Cu-anode tube, run at 60 kV - 40 mA; no filter, proportional counter (Xe), no pulse height selector.

ANSWER 16

The high background is due to the scatter by the specimen and to diffraction of all wavelengths contained in the primary spectrum against the set of strongly reflecting planes (d = 3.08 Å). The shape of this background should thus resemble the distribution of the white radiation, as no filter was used.

The Xe filled proportional counter shows a detector efficiency which drops sharply for wavelengths beyond the Xe absorption edge. The intensity, of the background will thus drop at an angular position corresponding to an energy of 34.5 keV.
Thus 34.5 keV corresponds to 0.359 Å.

$$2d\sin\theta = \lambda \text{ or } 2 \times 6.08 \times \sin\theta = 0.359$$

$$\therefore \theta = 3.365^\circ$$

$$\text{and } 2\theta = 6.73^\circ$$

The maximum intensity should be expected at approximately 25 keV, but the image of the white continuum should start at about 50 kV, or at a θ given by:

$$\theta = \sin^{-1} \frac{12.4}{50 \times 2 \times 3.08}$$

or a 2θ of about 4°.

This shape could not easily be confused with a diffraction line. However, the primary radiation is not only scatttered and diffracted by the specimen but also absorbed. The more radiation of the white continuum that is absorbed, the less intensity is scattered as background radiation.

As it happens, the K absorption edge of Ba is 37.2 keV. This means that radiation of 38 keV is more absorbed than higher energies. The scattered intensity should thus drop slowly going from low to high 2θ values; up to an angle corresponding to 37.2 keV.

This angle corresponds to:

$$\sin^{-1}\left(\frac{12.4}{37.2 \cdot 2 \cdot 3.08}\right) \text{ or } 6.20\ ^{\circ}(2\theta)$$

The intensity of the white spectrum still increases in this region, and so a more or less constant intensity is found. At 6.20° (2θ) however, a sudden increase in diffracted intensity is observed, as these wavelengths are not so strongly absorbed by the Ba.

The sudden increase between these two limits at 6.20° (2θ) and 6.75° (2θ) corresponding to 37.2 keV and 34.5 keV radiation scattered by the d = 3.08 Å planes, may well be confused with a regular diffraction line. This effect may be distinguished from other spurious peaks as they should only occur for compounds having a high concentration of elements like Cs, Ba, La and Ce with absorption edges not too far above that of Xe.

QUESTION 17

Counting statistics - variation of standard deviation with counting time

A A certain analysis line gave 800 c/s under certain instrumental conditions. Given that the standard deviation (σ) of a given number of counts (N) is equal to $\sqrt{N}$, calculate the standard deviation expected in a counting time of 20 seconds.

Assuming that the concentration of the compound giving that line was a %, what would the calculated standard deviation correspond to in terms of concentration.

B Given a counting rate of 150 c/s, prepare a curve of standard deviation (σ%) against analysis time, taking analysis times of 4, 10, 20, 40, 60 and 100 seconds.

ANSWER 17

A It is stated that $\sigma = \sqrt{N}$ and values of the counting rate (R) and analysis time (T) are given. Since $N = RT$ in the given example

$$N = 800 \times 20 = 16000$$

thus $\sigma = \sqrt{16000}$

$= \underline{127 \text{ counts}}$

In terms of percentage, $\sigma\,\% = \dfrac{\sqrt{N}\cdot 100}{N} = \dfrac{100}{\sqrt{N}}$

in the given example, $\sigma\,\% = \dfrac{100}{\sqrt{16000}} = 0.79\%.$

Since the given concentration is a %, the deviation will correspond to:

$$a \times \frac{0.79}{100} = \underline{0.0079\ a\ \%}$$

B Fig.10 shows a plot of standard deviation as a function of counting time.

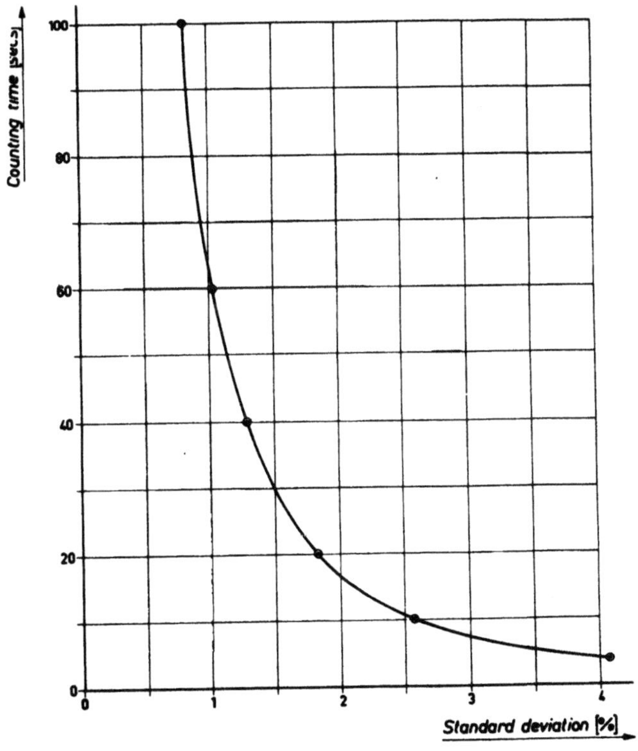

Fig.10

QUESTION 18

Estimation of error introduced by ignoring the background

Whenever X-ray intensity measurements are made on a selected peak, occuring at a certain setting of the goniometer, some fraction of the measured counting rate is always due to background. The significance of this background is dependent on its relative magnitude, and it is often difficult to judge whether or not it can be ignored in the calculation of the counting error. The true counting error ε % is related to the peak R_p and background R_b counting rates in the following way:

$$\varepsilon\% = \frac{100}{\sqrt{T}} \cdot \frac{1}{\sqrt{R_p} - \sqrt{R_b}} \tag{6}$$

where T is the total analysis time.

If the background is ignored, however, the counting error is calculated simply from:

$$\varepsilon\,\% = \frac{100}{\sqrt{N}} \tag{10}$$

where N is the total number of counts collected at the peak position.

Prepare a calibration curve which shows the error introduced by ignoring the background, for the following experimental conditions:

total analysis time 120 s

peak counting rate 100 c/s

The curve should cover the range of 2 - 25 in terms of peak to background ratio.

ANSWER 18

If the background were ignored all of the 120 seconds would be used in collecting counts on the peak (R_p). Thus the presumed error can then be calculated;

$$\varepsilon\,\% = \frac{100}{\sqrt{N}} \quad \text{and} \quad N = R \cdot T = 100 \times 120 = 12\,000$$

$$\varepsilon\,\% = 0.914$$

However the true error allowing for the background (R_b) can be estimated using the expression given in Equation 6.

$$\varepsilon\,\% = \frac{100}{\sqrt{T}} \cdot \frac{1}{\sqrt{R_p} - \sqrt{R_b}}$$

where T is the total analysis time. Since R_p is always 100 c/s the true error is given by:

$$\varepsilon(\%) = \frac{9.14}{10 - \sqrt{R_b}}$$

The attached calibration curve was constructed using the following data, see Fig.11.

R_b/R_p	R_b	$\sqrt{R_b}$	$\frac{9.14}{10-\sqrt{R_b}}$ %
2	50	7.07	3.11
3	33.3	5.77	2.18
4	25	5	1.83
6.25	16	4	1.52
10	10	3.16	1.34
14.3	7	2.64	1.24
20	5	2.24	1.18
25	4	2	1.14

The hatched area indicates the error introduced by ignoring the background.

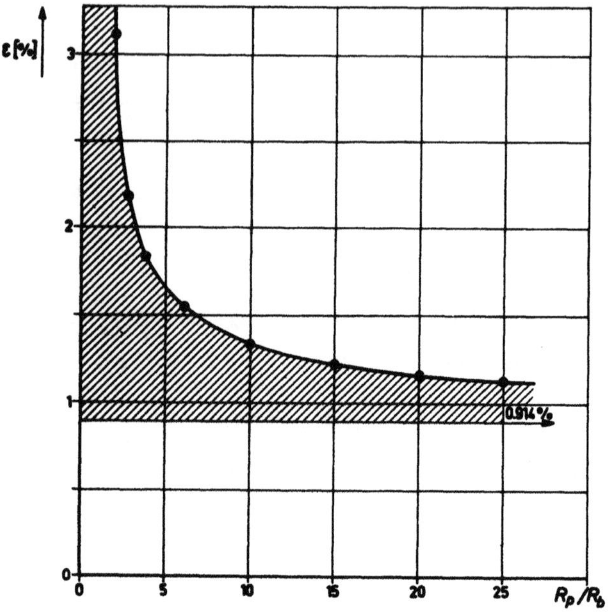

Fig. 11

QUESTION 19

Trace analysis - derivation of detection limit expression

A The lower limit of detection is often defined as that concentration equal to two standard deviations of the background count rate. Derive an expression which defines the lower limit of detection in terms of the analysis time T, the background counting rate R_b and the response m of an element given in c/s per %. Remember that trace analysis always requires two measurements, i.e. peak plus background.

B 0.3% of Cu in a mineral sample gave 464 c/s on the Cu Kα peak position and a background of 28 c/s. Calculate the lower limit of detection for a total analysis time of 40 seconds.

ANSWER 19

A An X-ray signal is said to be significantly greater than the background if

$$(R_p - R_b) \geqslant 2\sigma(\bar{R}_b)$$

$\bar{R}_b$ is the true (or mean) background value and $\sigma(R_b)$ is the standard deviation of the background measurement. In X-ray analysis neither $\bar{R}_b$ nor $\sigma(R_b)$ are known; it is necessary therefore, to measure R_b in time T_b and to use the result N_b/T_b as being equivalent to $\bar{R}_b$.

$$2\sigma(R_b) = \frac{\sigma N_b}{T_b} = \frac{\sqrt{R_b T_b}}{T_b} = \sqrt{\frac{R_b}{T_b}}$$

Thus at the detection limit:

$$(R_p - R_b) \geqslant 2\sigma(R_b) \quad \text{or} \quad \geqslant 2\sqrt{\frac{R_b}{T_b}}$$

To convert into concentration it is necessary to divide by m i.e. c/s/%. Thus the lower limit of detection (L.L.D.) will be given by

$$L.L.D. = \frac{2}{m}\sqrt{\frac{R_b}{T_b}}$$

Since, however, T_b represents only half of the total analysis time it is necessary to reduce further the expression by $\sqrt{2}$, giving a final expression

$$L.L.D. = \frac{2\sqrt{2}}{m}\sqrt{\frac{R_b}{T}}$$

where T represents the total analysis time.

B It is stated that 0.3% of Cu gives an R_p of 464 c/s and an R_b of 28 c/s.

Thus the value of m will equal $\frac{464 - 28}{0.3}$ = 1453 c/s/%

Substitution in the equation gives:

$$L.L.D. = \frac{2\sqrt{2}}{1453}\sqrt{\frac{28}{40}} = 1.6 \cdot 10^{-3}\ \%$$

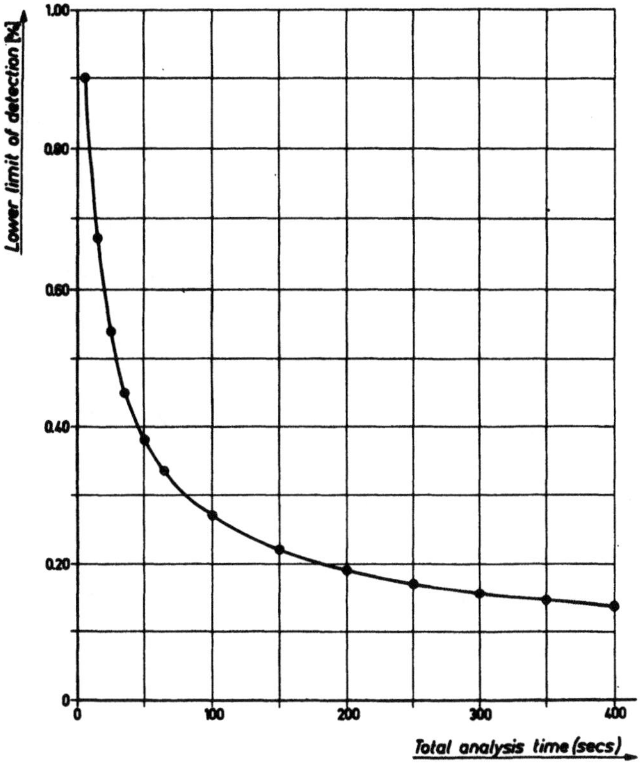

Lower limit of detection [%]
1.00
0.80
0.60
0.40
0.20
0
0
100
200
300
400
Total analysis time (secs)

Fig. 12

QUESTION 20

Choice between fixed time and fixed count

A series of samples of alloyed steels were found to give 300 c/s/% for the Si $K\alpha$ line. A calibration procedure has to be prepared over the range 0.20 - 0.60% Si and the choice has to be made between fixed count and fixed time operation.

A counting accuracy corresponding to 0.60 ± 0.02 % (2σ) Si has to be obtained at the upper end of the calibration range. Compare counting times and accuracies for the lower end of the calibration range, i.e. 0.20% Si, for the fixed count and fixed time method.

Note: The background counting rate can be assumed to be negligible.

ANSWER 20

Since the counting accuracy at the 0.60% concentration level has already been defined, it is possible to calculate the number of counts (*N*) and the analysis time (*T*) required.

At the 2σ confidence level $2\sigma(\%) = \dfrac{200}{\sqrt{N}}$

$0.60 \pm 0.02\,\%$ corresponds to a relative error of 3.33%, so

$$2\sigma = \frac{200}{\sqrt{N}} = 3.33 \text{ and } \underline{N = 3608 \text{ counts}}$$

It is stated that 300 c/s/% are obtained on the Si Kα line; so 0.60 % Si would give a counting rate (*R*) of 180 c/s.

Since $N = R \cdot T$ the analysis time will equal $\dfrac{3608}{180} = \underline{20.04 \text{ s}}$

At the low end of the concentration range there are two possibilities:

1. Fixed time: analysis time = 20.04 s i.e. same as for 0.60 % Si
   counting rate = 0.2 x 300 = 60 c/s
   number of counts = 60 x 20.04 = 1202 cts
   $$2\sigma(\%) = \frac{200}{\sqrt{1202}} = 5.77\% \text{ or a Si content of } 0.20 \pm 0.012\,\%$$
   in an analysis time of 20.04 s.

2. Fixed count: number of counts = 3608 (i.e. same as for 0.60 % Si)
   counting rate = 0.2 x 300 = 60 c/s
   analysis time $= \dfrac{3608}{60} = 60.13$ s.
   $2\sigma(\%) = 3.33$ (i.e. same as for 0.60 % Si)
   or a Si content of $0.20 \pm 0.007\,\%$ in an analysis time of 60.13 s.

In this instance, then the fixed time method is better since the slightly lower absolute error is preferable to a three times longer counting time at the low concentration end of the range.

QUESTION 21

Choice between ratio and absolute counting methods

The choice has to be made between the ratio and absolute method of counting in a counting programme which already involves collection of counts on a standard sample (i.e. whether the absolute or ratio method is chosen the standard is counted anyway, so no time is to be saved by not counting the standard).

The standard deviation of the equipment is known to be 0.2% in a ratio measurement, but $2\frac{1}{2}$ times worse in an absolute measurement.

Make the choice between absolute and ratio measurements for the following sets of data:

A peak counting rate on standard 10^5 c/s
peak counting rate on sample 8.1×10^4 c/s
counting time on standard and sample each 10 s.

B peak counting rate on standard 10^3 c/s
peak counting rate on sample 8.1×10^2 c/s
counting time on standard and sample each 10 s.

ANSWER 21

The overall error in an analysis ε_T is related to the equipment error ε_e and counting error ε_c in the following way:

$$\varepsilon_T = \sqrt{\varepsilon_e^2 + \varepsilon_c^2} \tag{11}$$

The error in a ratio measurement ε_r between standard (counting rate R_s) and sample (counting rate R_x) in an analysis time divided equally between R_s and R_x is:

$$\varepsilon_T = \frac{100}{\sqrt{R_s T_s}} \sqrt{1 + \frac{R_s}{R_x}} \tag{12}$$

A Ratio measurement:

$$\varepsilon_r = \frac{100}{\sqrt{10^5 \times 10}} \cdot \sqrt{1 + \frac{10^5}{8.1 \times 10^4}} = 0.149\ \%$$

$$\varepsilon_e = 0.2\% \qquad \varepsilon_T = \sqrt{(0.2)^2 + (0.149)^2} = 0.25\% \quad \text{ratio}$$

Absolute measurement:

$$\varepsilon_a = \frac{100}{\sqrt{R_x T_x}} = \frac{100}{\sqrt{8.1 \times 10^5}} = 0.111\ \%$$

$$\varepsilon_e = 0.5\ \% \qquad \varepsilon_T = \sqrt{(0.5)^2 + (1.11)^2} = 1.22\ \% \quad \text{absolute}$$

Thus the ratio measurement is better

B Ratio measurement:

$$\varepsilon_r = \frac{100}{\sqrt{10^3 \times 10}} \cdot \sqrt{1 + \frac{10^3}{8.1 \times 10^2}} = 1.49\ \%$$

$$\varepsilon_e = 0.2\ \% \qquad \varepsilon_T = \sqrt{(0.2)^2 + (1.49)^2} = 1.51\ \% \quad \text{ratio}$$

Absolute measurement:

$$\varepsilon_a = \frac{100}{\sqrt{8.1 \times 10^2}} = 1.11\ \%$$

$$\varepsilon_e = 0.5\ \% \qquad \varepsilon_T = \sqrt{(0.5)^2 + (0.111)^2} = 0.51\ \% \quad \text{absolute}$$

Thus in this instance the absolute measurement is preferable.

QUESTION 22

Variation of limit of detection with analysis time

A series of measurements of fluorine in cement sample yielded a peak counting rate of 4 c/s/% for fluorine (F Kα) over a background of 6.5 c/s.

Using the given formula for lower limit of detection (*L.L.D.*), prepare a calibration curve showing the variation in detection limit as a function of analysis time, up to a total time of 400 s.

$$L.L.D. = \frac{3}{m}\sqrt{\frac{R_b}{T_b}} \qquad (13)$$

where m is in c/s/%, R_b the background counting rate in c/s and T_b the counting time on the background, i.e. one half the total analysis time.

ANSWER 22

The lower limit of detection (*L.L.D.*) is given by:

$$L.L.D. = \frac{3}{m}\sqrt{\frac{R_b}{T_b}}$$

also given are values of m and R_b. Since T_b is one half of the total analysis time T the following expression gives the variation of detection limit analysis time:

$$L.L.D. = \frac{3}{4}\sqrt{6.5}\sqrt{\frac{2}{T}} = \frac{2.70}{\sqrt{T}}$$

The calibration curve in Fig.12 was constructed from the following data:

T (s.)	$\sqrt{T}$	$L.L.D. = \frac{2.70}{\sqrt{T}}$ (%)
9	3	0.90
16	4	0.675
25	5	0.54
36	6	0.45
50	7.06	0.383
64	8	0.337
100	10	0.27
150	12.25	0.221
200	14.14	0.191
250	15.81	0.171
300	17.32	0.156
350	18.71	0.144
400	20	0.135

QUESTION 23

Analytical error due to counting statistics

A series of measurements are being made to determine the concentration of BaO in a mixture of BaO and $BaO.6Fe_2O_3$.
A slow scan has been made over a suitable BaO reflection and a total of 32060 counts were collected. Collection of counts in a suitable neighbouring part of the spectrum gave 20461 counts, as background, in the same time.
Assuming that the concentration of BaO was 11.65 %, calculate the error due to counting statistics in terms of % BaO. The results should be expressed in terms of 95 % (i.e. 2σ) confidence limits.

The standard deviation (σ) of a measurement involving a given number of peak counts (N_p) and a given number of background counts (N_b) is:

$$\sigma(\%) = 100\frac{\sqrt{N_p + N_b}}{N_p - N_b} \qquad (14)$$

ANSWER 23

It is stated that the standard deviation σ is given by:

$$\sigma(\%) = 100\frac{\sqrt{N_p + N_b}}{N_p - N_b}$$

Since it is also given that N_p is 32060 and N_b is 20461, σ can be easily calculated.

$$\sigma\,\% = 100\ \frac{\sqrt{32060 + 20461}}{32060\ -\ 20641} = 100\ \frac{\sqrt{52521}}{11419}$$

$\therefore$ Counting error = 2.01%

In terms of % BaO this will be $2.01 \times \frac{11.65}{100} = 0.23\%$

It is necessary, however, to quote the error in terms of 2σ ; the final result will therefore be (0.23 x 2) = 0.47 % BaO.

QUESTION 24

Particle statistics in X-ray diffractometry

Only those crystallites having the reflecting planes almost parallel to the specimen surface can contribute to a certain reflection. The intensity of the resulting diffraction is thus dependent on this number of crystallites. Intensities of different diffraction lines or between different specimens can only be compared if the number of particles contributing is the same fraction of the total number of particles. If this total number is too small to warrant a random distribution, an error in the intensity measurement is introduced[8]. This may occur for instance when the particles are too large.

A Calculate the relative standard deviation of an intensity measurement due to the particle statistics for particles having a cubic shape with dimensions of 1, 5 and 10 μm for the following experimental conditions: diffractometer with an irradiated sample volume of 10 x 10 x 0.1 mm^3; filling factor = 0.5; multiplicity of this set of planes 2, focus width 2 mm, take-off angle 6°, radius of diffractometer 17 cm; receiving slit width 10 mm; diffraction angle $\theta = 30^o$. Assume that diffraction only takes place at the correct Bragg angle θ. The effect of Soller blades and divergence slits can be ignored.

B What can be done to decrease this error.

ANSWER 24

The standard deviation is dependent on the number of crystallites contributing to this diffraction maximum N and is given by $\sqrt{N}$, assuming a normal distribution. The relative standard deviation is thus

$$\frac{100}{\sqrt{N}}\ \%$$

A If it is assumed that the dimension of the particles is a mm, the volume is then a^3 mm^3. The total number of particles can thus be taken to be

$$\frac{10}{a^3} \times \tfrac{1}{2} = \frac{5}{a^3}$$

The particles with the reflecting planes exactly parallel to the surface and those with a slight variation Δ are both in a correct position for diffraction; this can be taken as the angle under which the focus is seen from the specimen, e.g.

$$\frac{2 \times \sin 6^\circ}{170} = \frac{1}{850} \text{ radians.}$$

Plane and counterplane can both diffract and the multiplicity is thus 2, the fraction of the total number of particles which can diffract is therefore given by:

$$\frac{1}{850} \times \frac{2 \times 2}{\pi} \times \cos 30^\circ$$

This diffracted intensity is spread out over a cone with opening angle 2θ.

The cross-section of this cone at the radius r or at the receiving slit is thus

$$\frac{10}{2 \times \pi \times 170 \times \sin 60^\circ}$$

The number of particles giving a diffraction line in the exact direction θ registered by the counter is then:

$$N = \frac{5 \times 2 \times 2 \times \cos 30^o \times 10 \times 10^9}{a^3 \times 850 \times \pi \times 2\pi \times 170 \times \sin 60^o} = \frac{2 \times 10^9}{a^3 \times 286 \times 10^2} = \frac{70000}{a^3}$$

where a is expressed in μm.

The relative standard deviation is thus

$$\frac{100}{\sqrt{70000}} \sqrt{a^3}\ \% = \frac{1}{7} \sqrt{7a^3}\ \%$$

Thus for the data given:

dimensions (μm)	1	5	10
N	70000	560	70
relative standard deviation	0.4%	4.2%	12%

This error is in practice less as the crystals exhibit a rocking curve, e.g. a certain region over which reflection occurs. If this is taken to be 0.10 $^o\theta$, then the number of diffracting particles increases, as Δ increases by

$$\left\{\frac{1}{850} + \frac{0.10}{57.3}\right\} : \frac{1}{850}$$

or approximately by a factor of 2.5 times. This means that the error is reduced by a factor of about 1.6 times.

B Oscillating the specimen will again increase this variation Δ, but might lead to broader diffraction lines. Rotation of the specimen in its own plane will greatly enhance the probability of diffraction and therefore reduce the error. However, in case of preferred orientation e.g. where platelets are oriented parallel to the specimen surface, specimen rotation does not decrease the particle distribution error.

QUESTION 25

Determination of traces of lead in oil (absolute measurement)

A series of oil samples gave the following intensities on the Pb Lα line:

Sample	Scaler value (counts)	p.p.m. Pb
1	4104	10
2	4968	100
3	4423	45
4	4011	1
5	4025	4
6	4657	?

The following instrumental conditions were used:

tungsten anode X-ray tube, 40 kV, 20 mA; LiF crystal; scintillation counter; fixed time 32 s; scale factor 32.

Prepare a calibration curve and work out the equation of the line.

What is the concentration of lead in sample 6.

ANSWER 25

A calibration curve is constructed by plotting the reported concentration(abscissa)against the counting rate (ordinate). The counting rate is obtained by dividing the number of counts collected by the analysis time. The number of counts collected is the value read on the scalar multiplied by the scale factor, i.e.

$$\text{counting rate} = \frac{\text{number of counts}}{\text{time}} = \frac{\text{scaler value x scale factor}}{\text{time}}$$

Since in this instance the scale factor and the analysis time have the same numerical value, the scaler value is the same as the counting rate.

A calibration line is drawn by fitting a straight edge as closely as possible to minimize the distances for each of the points from the line (i.e. the best regression fit), see Fig.13.

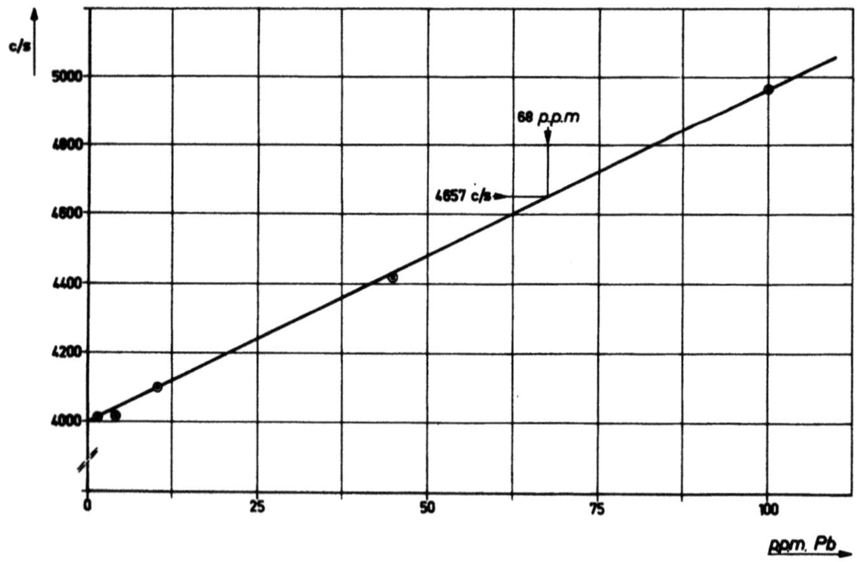

Fig. 13

Unknown concentrations can either be read straight from the calibration graph, or alternatively, calculated using the equation for the line derived from the general equation for a straight line:

$$\frac{Y - Y_1}{Y_2 - Y_1} = \frac{X - X_1}{X_2 - X_1} \qquad (15)$$

Select two sets of ordinates, e.g. Y_1, X_1 = 4000 c/s, 0 p.p.m.

Y_2, X_2 = 4968 c/s, 100 p.p.m.

Then substitute in the general equation:

$$\frac{X - 0}{100 - 0} = \frac{Y - 4000}{4968 - 4000}$$ (note: Y is counts and X = p.p.m. Pb)

thus: p.p.m. lead = (0.1034 x count rate) - 413

or for sample 6:

p.p.m. lead = (0.1034 x 4675) - 413

= 481 - 413 = 68 p.p.m.

QUESTION 26

Determination of CaO in cement (ratio measurement)

The following data were obtained using a series of cement samples. Counting ratios were determined on the Ca Kα line using sample 5 as the ratio standard.

sample	counting ratio	% CaO
1	0.959820	59.39
2	0.978200	?
3	0.986080	61.58
4	0.989340	?
5	1.000000	62.61
6	1.004910	63.11
7	1.051400	66.50

Equipment conditions were as follows:
chromium anode tube, 60 kV - 4 mA; PE crystal; flow counter; fixed count on standard of 10^6 cts.

Prepare a calibration curve and calculate the equation of the line. Determine the concentration of CaO in samples 2 and 4.

ANSWER 26

Draw the calibration curve by plotting concentration of CaO (abcissa) against count ratio (ordinate)*, see Fig. 14.

Select two sets of ordinates and determine the slope of the line, e.g.:

$$C_1, R_1 = 60\%, 0.9640$$

$$C_2, R_2 = 65\%, 1.0314$$

$$\frac{C - 60}{65 - 60} = \frac{R - 0.9640}{1.0314 - 0.9640}$$

$$\underline{C = 74.18 R - 11.51}$$

Calculate the concentration of CaO in samples 2 (C2) and 4 (C4).

$$C_2 = (74.18 \times 0.9782) - 11.51 = 61.05\%$$

$$C_4 = (74.18 \times 0.9893) - 11.51 = 61.87\%$$

*Note In the ratio measurement the counting rate ratio between sample R_x and standard R_s is required. Since count rate is number of counts, N, divided by time, T, the following expression holds:

$$\frac{R_x}{R_s} = \frac{N_x}{T_x} \cdot \frac{T_s}{N_s}$$

A number of counts, N_s, is selected - in this case 10^6 - and the time required to collect these counts, T_s, is stored in timer unit. A number of counts, N_x, is then collected in the same time, T_x, where $T_s = T_x \cdot N_x$ is then printed out, this being the counting ratio since:

$$\frac{R_x}{R_s} = \frac{N_x}{\not{T_x}} \cdot \frac{\not{T_s}}{N_s}$$

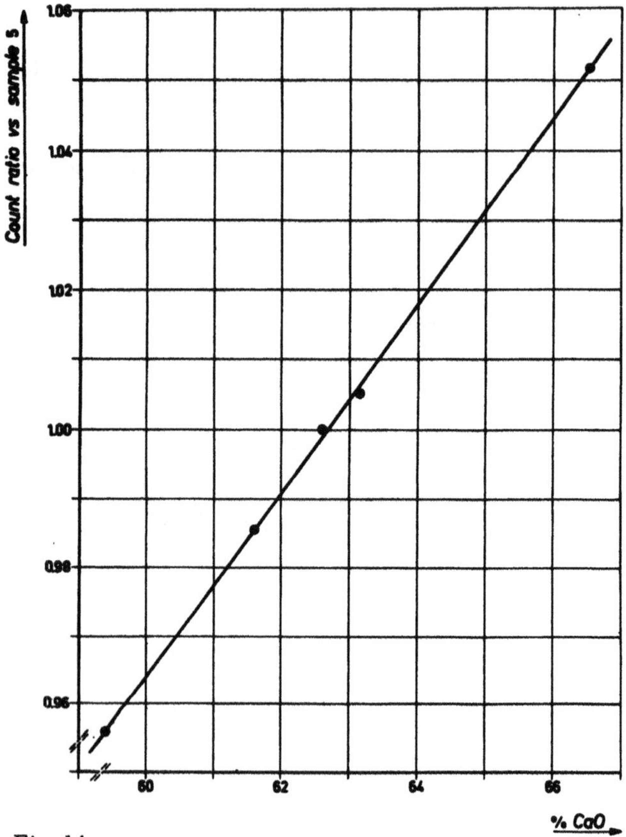

Fig. 14

QUESTION 27

Determination of nickel in steel (with dead time correction)

A series of steel samples yielded the following count data:

sample	scaler time (s)	% Ni
1	20.8	7.90
2	34.3	4.00
3	86.8	1.39
4	30.5	4.71
5	24.9	?
6	41.4	3.22

The following equipment conditions were employed:

tungsten anode tube, 40 kV - 24 mA; quartz crystal; Geiger-Müller counter (dead time 240 μs); fixed counts 25600.

Prepare calibration curves with and without the use of a dead time correction and determine the concentration of nickel in sample 5.

ANSWER 27

It is stated that the fixed number of counts collected is 25600 and the analysis times are given. Thus by dividing 25600 by the recorded time the measured counting rate is obtained in c/s. A plot of counting rate against reported nickel concentration yields the uncorrected curve shown in Fig.15. It is, however, immediately noticeable that the curve is very curred due to the effect of dead time.

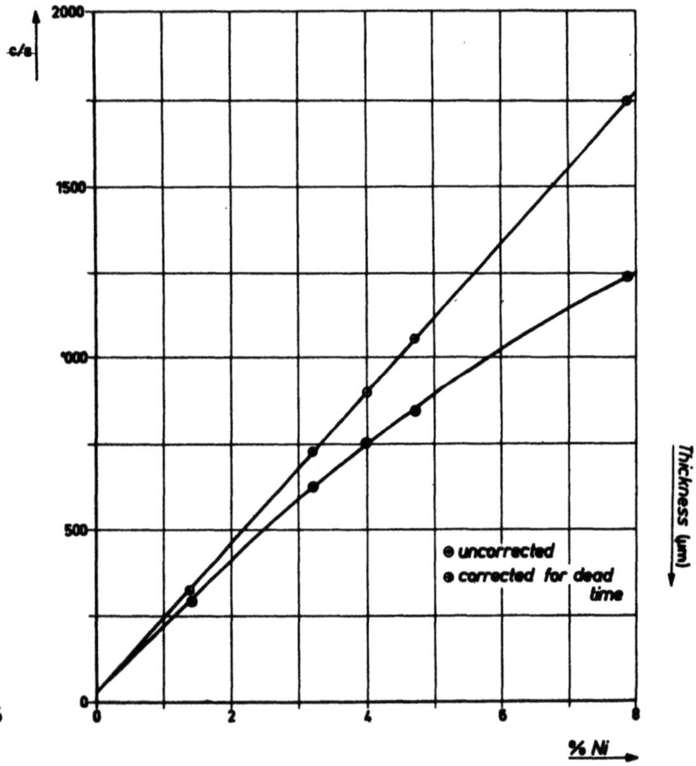

Fig.15

One of the advantages of working with the fixed count method is that the absolute time lost due to dead time is always constant and in fact equal to the product of the dead time and the number of counts taken. Thus in the quoted example the absolute time loss is always $(25600 \times 240 \times 10^{-6})$ s

= 6.14 s. The true counting rate is determined not by dividing the number of counts collected by the time indicated on the timer (the apparent time), but rather by the apparent time, less the lost time interval (the true time) thus:

sample	timer indication s	true time s	true counting rate (c/s)	% Ni
1	20.8	14.66	1748	7.90
2	34.3	28.16	909	4.00
3	86.8	80.66	317	1.39
4	30.5	24.36	1052	4.71
5	24.9	18.76	1365	?
6	41.4	35.26	726	3.22

The concentration of nickel in sample 5 can either be read from the curve or, as in the previous example, by use of the equation for the line.

The equation for the line is $C = 0.00457I - 0.100$, where C is the nickel concentration in % and I the true count rate.

The value of the concentration of nickel in sample 5 is 6.14%.

QUESTION 28

Calculation of matrix absorption and prediction of absorption effects

The total secondary absorption of a matrix is given by

$$\mu_{matrix} = \sum_i (\mu_i w_i) \tag{16}$$

where μ_i and w_i are the individual mass absorption coefficients and weight fractions for each element i making up the matrix.

The values of the mass absorption coefficients for Ni and Cr for Ni Kα radiation are 61 and 316 respectively.

A What is the total secondary absorption for Ni Kα radiation of a nichrome alloy with the composition 40% Cr, 60% Ni

B The slope of a calibration curve (Counting rate for a certain wavelength against concentration of the element giving that wavelength) is to a first approximation inversely proportional to the total secondary absorption.
It is necessary to prepare a calibration curve for Ni in a series of nichrome alloys containing a range of Cr from 60% to 80%. Calculate the range of secondary absorption and predict what effect this is likely to have on the calibration curve.

ANSWER 28

It is given that the total matrix absorption is the sum of the products of each individual mass absorption coefficient and its weight fraction, i.e.

$$\mu_{matrix} = \sum_i (\mu_i w_i)$$

A In the given example

$$(\mu_{Ni/Cr})^{NiK\alpha} = (\mu_{Ni} \times w_{Ni}) + (\mu_{Cr} \times w_{Cr})$$

$$(\mu_{60/40})^{NiK\alpha} = (61 \times 0.6) + (316 \times 0.4) = 163$$

B Since the range of Cr concentration is 60 to 80% the range of the absorption can be estimated by calculating the matrix absorption for 40Ni/60Cr and 20Ni/80Cr, i.e.

$$(\mu_{40/60})^{NiK\alpha} = (61 \times 0.4) + (316 \times 0.6)$$
$$= 24.4 + 189.6 = 214$$
$$(\mu_{20/80})^{NiK\alpha} = (61 \times 0.2) + (316 \times 0.8)$$
$$= 12.2 + 252.8 = 265$$

∴ the range in absorption is 214 (60 % Cr) to 265 (80 % Cr).

Since it is stated that the slope of a calibration curve is inversely proportional to the absorption, the curve between 60 and 80 % Cr would be expected to have a smooth decreasing slope, the total change in slope corresponding to 214/265 = 0.81.

QUESTION 29

Calculation of absorption coefficients and absorption effects

A Calculate the total secondary absorption of a cement sample for Ca $K\alpha$ radiation. Use the following mass absorption coefficients for Ca $K\alpha$.

	μ (Ca $K\alpha$)	composition of cement	
Ca	141	CaO	68%
Si	556	SiO_2	23%
Al	450	Al_2O_3	6%
Fe	318	Fe_2O_3	3%
O	115		

B The mass absorption coefficient of potassium for Ca $K\alpha$ radiation is 1080. Calculate the change in total absorption assuming that 2% of K_2O is added to the cement.

ANSWER 29

A It is first necessary to calculate the elemental composition of the cement

68% CaO	or	68 x 40.1/ 56.1 = 48.6% Ca
		68 x 16 / 56.1 = 19.4% O
23% SiO_2	or	23 x 28.1/ 60.1 = 10.8% Si
		23 x 32 / 60.1 = 12.2% O
6% $Al_2/_3$	or	6 x 54 /102 = 3.2% Al
		6 x 48 /102 = 2.8% O
3% Fe_2O_3	or	3 x 117 /159.7 = 2.1% Fe
		3 x 48 /158.7 = 0.9% O
		total oxygen = 35.3%

Since the total absorption is equal to $\Sigma(\mu_i w_i)$, in this instance:

$$\mu_{total} = (141 \times 0.486)^{Ca} + (556 \times 0.108)^{Si} + (450 \times 0.031)^{Al} + (318 \times 0.021)^{Fe} + (115 \times 0.353)^{O}$$

$$= 68.5 + 60.1 + 14.4 + 6.7 + 40.6 \therefore \mu_{total} = \underline{190.3}$$

B The effect of adding 2% of K_2O can be calculated as follows:

2% K_2O is equivalent to 2 x 78.2/94.2 = 1.66% K

2 x 16 /94.2 = 0.34% O

The new absorption will now equal (old absorption x 0.98) + absorption due to K_2O

$$\text{or } \mu_{total} = (190.3 \times 0.98)^{\text{old matrix}} + (1080 \times 0.0166)^{K} + (115 \times 0.0034)^{O}$$

$$= 186.5 + 17.9 + 3.9$$

$$\underline{\mu_{total} = 208.3}$$

Thus the relative change in $\mu = \dfrac{208.3 - 190.3}{190.3} = 9.5\%$. This would lead to a Ca $K\alpha$ intensity being too low by 9.5% or to a concentration for Ca too low by 9.5%, assuming a linear calibration slope passing through zero intensity at zero concentration.

QUESTION 30

Absorption correction involving primary and secondary absorption

The slope of a calibration curve for a certain element is dependent upon the total absorption of the measured wavelength within the sample matrix. Although it is common practice to assume that only secondary absorption is of significance, in fact both primary and secondary absorption should be considered. The following expression relates the so-called "efficiency factor" $C(\lambda\lambda_j)$ with the mass absorption coefficient μ_i of each element i for the measured wavelength λ_j and an effective primary wavelength λ which represents the primary spectrum.

W_i is the weight fraction of each individual matrix element. Thus:

$$C(\lambda\lambda_j) = \frac{\mu_j(\lambda)}{\sum_i W_i\left(\mu_i(\lambda) + A\mu_i(\lambda_j)\right)} \tag{17}$$

A is a geometric factor having a value of 1.5.

Assuming that the slope of a calibration curve is directly proportional to the efficiency factor, calculate the likely variation in the slope of Fe $K\alpha$ in a series of binary alloys of Cr and Fe in which the Cr varies from 5 to 25%.

A value of 1.25 Å can be taken for the effective wavelength λ and the following values should be taken for the mass absorption coefficients:

	μFe	μCr
effective wavelength (1.25 Å)	161	131
Fe $K\alpha$ (1.937 Å)	71	445

ANSWER 30

In order to estimate the likely variation in the calibration curve for Fe Kα, it is first necessary to calculate the absorption values for the two extremes of the calibration range, i.e. 5% and 25% Cr.

The general form of $C(\lambda\lambda_j)$ is given as:

$$C(\lambda\lambda_j) = \frac{\mu_j(\lambda)}{\Sigma W_i\left(\mu_i(\lambda) + A\mu_i(\lambda_j)\right)}$$

or in the given case:

$$\left(C(\lambda\lambda_j)\right)^{FeK\alpha} = \frac{(\mu)_{Fe}^{1.25}}{W_{Fe}\left((\mu)_{Fe}^{1.25} + 1.5\ (\mu)_{Fe}^{FeK\alpha}\right) + W_{Cr}\left((\mu)_{Cr}^{1.25} + 1.5\ (\mu)_{Cr}^{FeK\alpha}\right)}$$

$$= \frac{161}{W_{Fe}\ (161 + 1.5 \times 71) + W_{Cr}\ (131 + 1.5 \times 445)}$$

$$= \frac{161}{267W_{Fe} + 799W_{Cr}}$$

Thus for 5%Cr $\quad \left(C(\lambda\lambda_j)\right)_{5\%}^{FeK\alpha} = \dfrac{161}{(267 \times 0.95) + (799 \times 0.05)} = \dfrac{161}{294}$

and for 25%Cr $\quad \left(C(\lambda\lambda_j)\right)_{25\%}^{FeK\alpha} = \dfrac{161}{(267 \times 0.75) + (799 \times 0.25)} = \dfrac{161}{401}$

It is thus apparent that the matrix absorption increases in the ratio 401/294 over the range 5% Cr to 25% Cr. Thus the calibration curve for Fe Kα over the equivalent range i.e. 75% to 95%, will decrease in the same ratio giving a calibration curve of increasing slope.

QUESTION 31

Calculation of quantity of heavy absorber needed to minimize matrix effects

In the analysis of tin ores it is found that a very curved calibration line is obtained due to the large difference in the mass absorption coefficients of the two matrix components Sn (μ = 13) and SiO_2 (μ = 1.80) for the analysis line (Sn Kα).

Calculate how much BaO heavy absorber (μ = 17) must be added to reduce the deviation from linearity to 1% over the range 0-20% Sn. It can be assumed that the slope of the calibration curve is inversely proportional to the matrix mass absorption coefficient. Only secondary absorption need be considered in this instance.

ANSWER 31

It is given that the change in the slope is inversely proportional to the change in the mass absorption coefficient μ. Since the maximum allowable deviation from linearity after dilution is 1%, this in turn requires that the difference in the μ values of the diluted samples must also equal 1%.

The total mass absorption coefficient of a matrix $\mu_{matrix} = \sum_i (\mu_i w_i)$ where μ_i and w_i are respectively the individual μ and weight fraction values of each element in the matrix. After dilution, the matrix μ of the 0% Sn sample (μ_0) and of the 20% Sn sample (μ_{20}) are made up as follows (the weight fraction of the added BaO is taken as x):

		Sn		SiO_2		BaO		
μ_0	=	$0\,(1-x)\,13$	+	$1\,(1-x)\,1.8$	+	$(x)\,17$	=	$15.2\,x + 1.8$
μ_{20}	=	$0.2\,(1-x)\,13$		$0.8\,(1-x)\,1.8$	+	$(x)\,17$	=	$12.96\,x + 4.04$

In each case the figures in parentheses represent the weight fractions of sample (ore) and diluent (BaO). In the case of the ore the weight fraction $(1-x)$ has to be distributed between Sn and SiO_2 in the ratio 0 : 1 for μ_0 and 0.2 : 0.8 for μ_{20}.

Since Sn has a higher mass absorption coefficient than SiO_2 it follows that the slope at 20% Sn (m_{20}) is less than that at 0% Sn (m_0). Since μ is inversely proportional to m this means that μ_{20} is greater than μ_0 and in fact, as stated originally, the two differ by 1%; i.e.,

$$\mu_{20} = 1.01\,\mu_0$$

Substitution of the values of μ_{20} and μ_0 allow the solution for x;

$$12.96\,x + 4.04 = 1.01\,(15.2\,x + 1.08)$$

$$\text{or } x = 0.929$$

In other words, the sample to BaO ratio should be 0.071 : 0.929 or a dilution ratio of about 1 : 13.

QUESTION 32

Use of α correction factor (intensity correction)

A series of eight samples were analysed for a certain element A. An element B, also present in major concentration, was found to strongly absorb element A. The following intensity data were obtained:

sample no.	intensity element A(c/s)	concentration element A(%)	intensity element B(c/s)
1	41	12.5	303
2	83	25.0	292
3	128	37.5	265
4	178	50.0	229
5	236	62.5	187
7	401	87.5	51
6	311	75.0	117
8	500	100.0	0
X	139	?	143
Y	356	?	87

The concentrations of element B are unknown, although it is suspected that the eight standard samples are almost pure binary mixtures.

Prepare a calibration curve of intensity versus concentration of element A.

It can be shown that it is possible to correct for the effect of element B using the following expression:

$$C_A = \frac{I_A}{m_A}\left(1 + K_{AB} I_B\right) \tag{18}$$

where C_A is the concentration of element A giving intensity I_A and m_A is the c/s/% for element A in the absence of element B. I_B is the intensity of element B. K_{AB} is the so-called "α" correction factor representing the effect of B on A.

Determine the value of K_{AB} and replot the curve for element A using corrected intensities (i.e. corrected for the effect of B). Determine the concentration of A in samples X and Y.

ANSWER 32

The following expression is given:

$$C_A = \frac{I_A}{m_A}\left(1 + K_{AB} I_B\right)$$

If this is rewritten in the form $y = mx + c$ the following is obtained:

$$\frac{m_A\, C_A}{I_A} = K_{AB} I_B + 1$$

So by plotting $\frac{m_A\, C_A}{I_A}$ on the ordinate and I_B on the abcissa a curve should be obtained, the slope of which is K_{AB} and which cuts the ordinate at unity, where $I_B = 0$. Thus, since $m_A = \frac{500}{100} = 5$.

sample no.	C_A/I_A	$m_A \cdot C_A/I_A$	$0.00173\ I_B$	$I_A\ (1 + 0.00173\ I_B)$
1	0.305	1.525	0.524	62.6
2	0.301	1.505	0.506	125
3	0.291	1.455	0.458	187
4	0.281	1.405	0.397	249
5	0.265	1.325	0.324	313
6	0.241	1.205	0.202	374
7	0.218	1.090	0.088	437
8	0.200	1.000	0	500

Plotting these values as in Fig. 16 gives a value of K_{AB} = 0.00173. Fig. 17

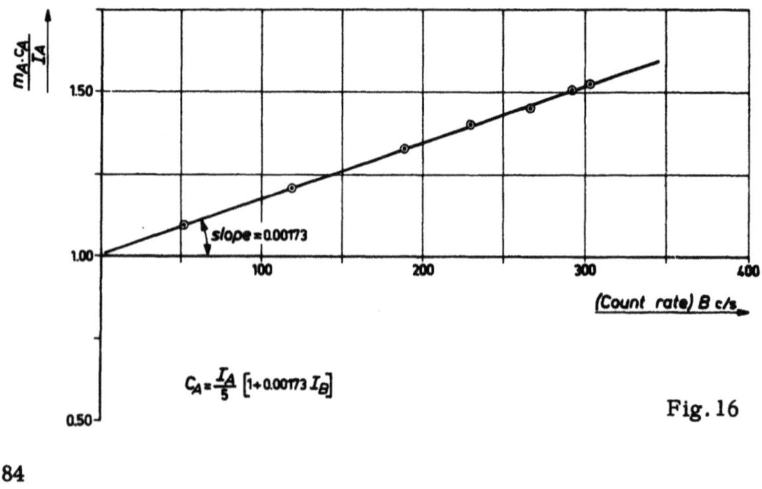

Fig. 16

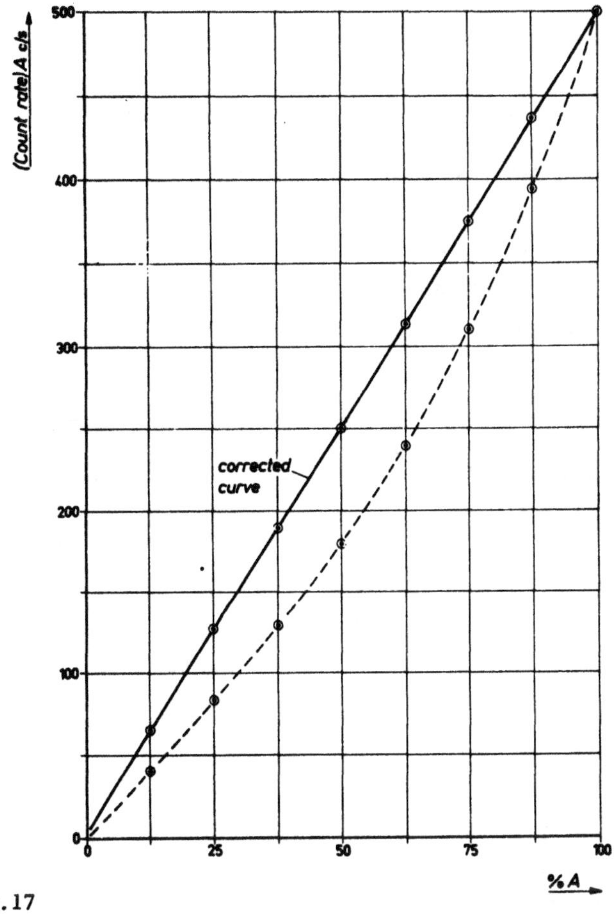

Fig. 17

shows the calibration curve of concentration versus intensity for element A.

Intensities are corrected by multiplying by $(1 + 0.00173 I_B)$ and when replotted a straight line is obtained as also shown in Fig. 17.

Concentrations of element A in samples X and Y are as follows:

sample X: $C_A = \frac{139}{5} [1 + (0.00173 \times 143)] = 27.8 \times 1.249 = 34.7\%$

sample Y: $C_A = \frac{356}{5} [1 + (0.00173 \times 87)] = 71.2 \times 1.151 = 81.9\%$

QUESTION 33

Use of α correction factors (concentration correction)

A series of samples are being analysed for lead using the Pb Lα line. It is found that zinc which is also present strongly absorbs the Pb Lα wavelength resulting in a very poor calibration curve. The following data were obtained:

sample nr.	% lead	Pb Lα (c/s)	% zinc
1	0.03	40.6	2.97
2	0.03	46.7	0.53
3	0.10	155.3	0.545
4	0.11	147.4	3.12
5	0.30	466.9	0.54
6	0.31	421.0	2.92
7	0.18	256.5	1.92
8	0.24	360.0	1.07
X	?	310.2	0.86
Y	?	446.7	2.12

Application of the intensity correction method described in the previous example was found to yield rather poor results because Zn itself is strongly absorbed by a third unidentified element. Fortunately Zn concentrations are available so a correction based on changes in the Zn concentration can be applied.

Plot a calibration curve of Pb Lα (c/s) versus % lead. Using the following expression and the data above calculate the correction factor K_{PbZn} representing the effect of Zn on Pb.

$$C_{Pb}=\frac{I_{Pb}}{m_{Pb}}\left(1+K_{PbZn}C_{Zn}\right) \tag{19}$$

C_{Pb} is the concentration of lead giving intensity I_{Pb}. m_{Pb} is the value of c/s/% for lead in the absence of interference from zinc. C_{Zn} is the concentration of zinc. (Note m_{Pb} can be estimated by plotting c/s/% for lead versus concentration zinc and extrapolating to zero % zinc).

Use the zinc correction factor to correct the lead intensity to zero % zinc and replot the calibration curve.

Finally use Equation (19) to determine the concentration of lead in samples X and Y.

ANSWER 33

Fig.18 shows the plot of Pb Lα (c/s) versus % Pb. Fig.19 shows a plot of Zn concentration against Zn intensity and indeed confirms the poor correlation due to the effect of the unidentified element.

In order to calculate the factor K_{PbZn} it is first necessary to estimate the number of c/s/% for Pb Lα, i.e. m_{Pb}, without influence from zinc.

sample nr.	(c/s/%) Pb Lα	% Zn	$\frac{C_{Pb} \cdot m_{Pb}}{I_{Pb}}$
1	1353	2.97	1.188
2	1557	0.53	1.037
3	1553	0.545	1.034
4	1340	3.12	1.194
5	1556	0.54	1.036
6	1358	2.92	1.192
7	1432	1.92	1.117
8	1510	1.07	1.060

Fig.20 shows a plot of (c/s/%) Pb Lα versus % Zn. Extrapolating the curve to zero % zinc gives a value of 1600 c/s/% for m_{Pb}.

Rearrangement of Equation (19) gives:

$$\frac{C_{Pb} \cdot m_{Pb}}{I_{Pb}} = K_{PbZn} \cdot C_{Zn} + 1$$

Fig. 21 shows a plot of $\frac{C_{Pb} \cdot m_{Pb}}{I_{Pb}}$ (ordinate) versus C_{Zn} (abcissa).

The slope of this line is K_{PbZn} and a value of 0.0637 is obtained. Thus the working equation for determining the concentration of lead is

$$C_{Pb} = \frac{I_{Pb}}{1600}\left(1 + 0.0637 C_{Zn}\right)$$

It is possible to correct the lead intensities for the effect of zinc by multiplying by the factor $(1 + 0.0637\ C_{Zn})$; so

sample nr.	Pb Lα (c/s)	$(1 + 0.0637\ C_{Zn})$	Pb Lα (c/s) corrected
1	40.6	1.189	48.3
2	46.7	1.034	48.3
3	155.3	1.035	160.7
4	147.4	1.198	176.9
5	466.9	1.034	482.4
6	421.0	1.186	499.0
7	256.5	1.122	288.3
8	360.0	1.068	384.6

Fig.18 shows a replot of the Pb Lα intensity data after correction.

Equation (19) can be used to calculate the concentration of lead in samples X and Y.

Sample X: $$C_{Pb} = \frac{310.2}{1600}\left[1 + (0.0637 \times 0.86)\right]$$

$$= 0.1938 \times 1.055 = 0.205\%$$

Sample Y: $$C_{Pb} = \frac{446.7}{1600}\left[1 + (0.0637 \times 2.12)\right]$$

$$= 0.279 \times 1.135 = 0.317\%$$

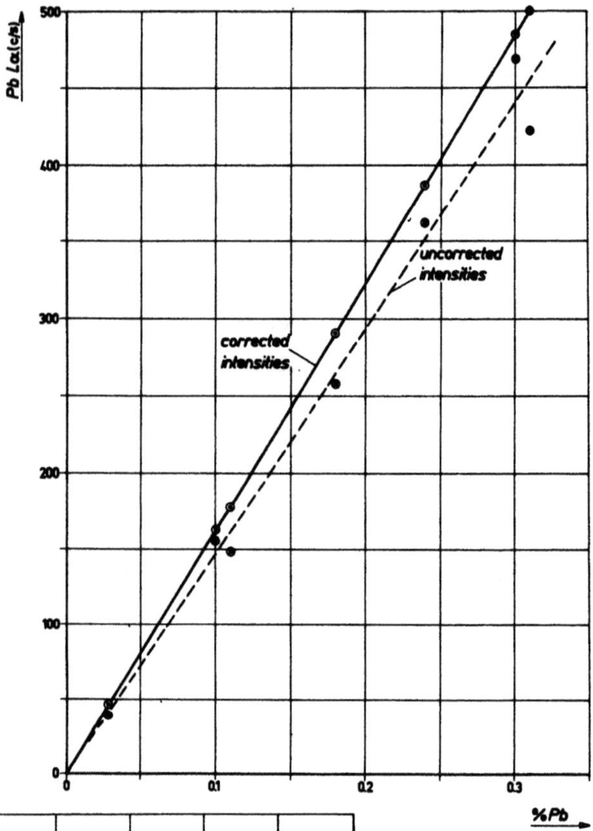

Pb Lα (c/s)
500
400
300
200
100
0
0
0.1
0.2
0.3
%Pb
corrected intensities
uncorrected intensities

Fig. 18

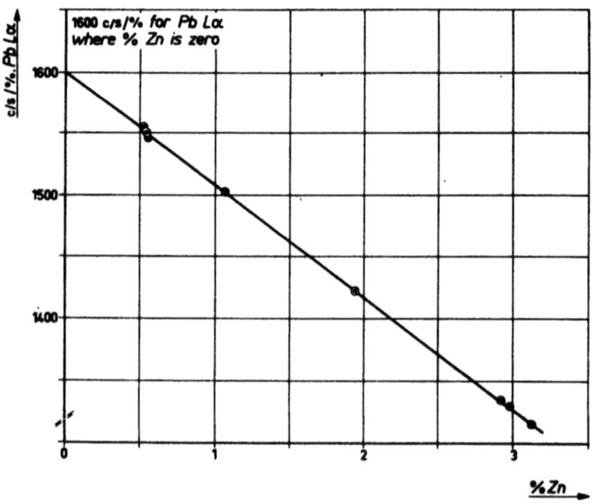

c/s /%, Pb Lα
1600 c/s/% for Pb Lα
where % Zn is zero
1600
1500
1400
0
1
2
3
%Zn

Fig. 20

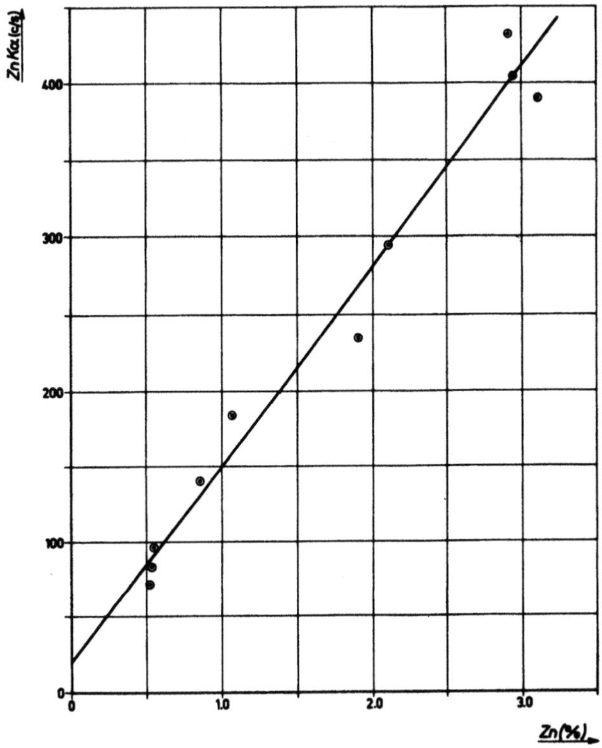
Zn Kα (c/s)
400
300
200
100
0
0
1.0
2.0
3.0
Zn (%)

Fig. 19

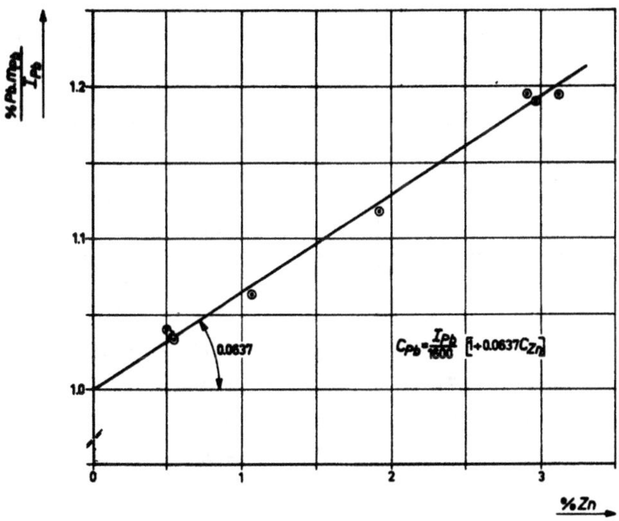
% Pb. mPb / IPb
1.2
1.1
1.0
0
1
2
3
0.0637
CPb = IPb / 1600 [1 + 0.0637 CZn]
% Zn

Fig. 21

QUESTION 34

Excitation probability

It is often thought that the wavelength in the primary spectrum most efficient in exciting the fluorescent radiation of an element A is that which is closest to the absorption edge of element A. The probability function of the primary spectrum should follow the absorption curve for A.

The probability of a wavelength in the primary spectrum giving a secondary photon can be expressed by:

$$W(\lambda) = \frac{\mu_A(\lambda)}{\Sigma \mu_i(\lambda) w_i + A \Sigma \mu_i(\lambda_A) w_i} \tag{20}$$

where $\mu_A(\lambda)$ is the mass absorption coefficient of A for λ

$\mu_i(\lambda)$ and $\mu_i(\lambda_A)$ are mass absorption coefficients of i for primary and λ_A.

w_i are weight fractions

A is a geometric factor and can be taken to be 1.5.

Check this assumption for aluminium by plotting μ (λ)and; w (λ) for pure aluminium, 1% Al in 99% Si and 20% Al in 80% Si. Is any matrix effect expected in these latter two cases? The following approximate absorption coefficients can be used:

λ (Å)	μ_{Al}	μ_{Si}	λ (Å)	μ_{Al}	μ_{Si}
5.00	1300	1600	6.50	2550	3100
5.25	1500	1800	6.75	2800	3500
5.50	1700	2000	7.00	3100	305
5.75	1900	2250	7.25	3400	325
6.00	2100	2500	7.50	3750	360
6.25	2300	2800	7.75	4100	395

μ (Al) for 8.34 Å = 330, μ (Si) for 8.34 Å = 480

ANSWER 34

The curve for μ (Al) can be plotted and shows the expected shape, see Fig.22. The results for the calculation of W(Al) are collected in the following table.

λ (Å)	Pure Al			1%Al/99%Si			20%Al/80%Si		
	μ_p	$\mu_p+1.5\mu_{Al}$	$W(\lambda)$	μ_p	$\mu_p+1.5\mu_{Al}$	$W(\lambda)$	μ_p	$\mu_p+1.5\mu_{Al}$	$W(\lambda)$
5.00	1300	1800	0.72	1597	2317	0.56	1530	2205	0.59
5.25	1500	2000	0.75	1797	2517	0.60	1740	2415	0.62
5.50	1700	2200	0.77	1997	2717	0.63	1940	2615	0.65
5.75	1900	2400	0.79	2247	2967	0.64	2160	2835	0.67
6.00	2100	2600	0.81	2496	3216	0.65	2420	3095	0.68
6.25	2300	2800	0.82	2795	3515	0.66	2700	3375	0.68
6.50	2550	3050	0.84	3094	3814	0.67	2990	3665	0.70
6.75	2800	3300	0.85	3490	4210	0.67	3360	4035	0.70
7.00	3100	3600	0.86	333	1053	2.94	865	1540	2.05
7.25	3400	3900	0.87	356	1076	3.14	920	1595	2.13
7.50	3750	4250	0.88	392	1112	3.36	1040	1715	2.18
7.75	4100	4600	0.89	432	1152	3.56	1135	1810	2.26
	$1.5\,\mu\,(Al) = 1.5 \times 330 = 500$			$1.5(0.99 \times 480 + 0.01 \times 330) = 720$			$1.5(0.80 \times 480 + 0.20 \times 330) = 675$		

The resulting graphs show clearly that $W(\lambda)$ does not follow the mass absorption coefficient either for pure Al or for the hypothetical mixture Al/Si. It is to be expected that $W(\lambda)$ will follow μ (λ) only if the denominator $\sum_i \mu_i(\lambda) w_i + A \sum_i \mu_i(\lambda_A) w_i$ is independent of λ.

$W(\lambda)$ varies only slightly between 5.00 and 6.74 Å for Al and Al/Si. There is very little difference between 1%Al/99%Si and 20%Al/80%Si. Beyond the Si-absorption edge, however, there is a tremendous difference in $W(\lambda)$. This means that almost no matrix effects are to be expected in the Al/Si system if a primary spectrum is used which contains no wavelengths longer than 6.74 Å. However, if radiation is used with wavelengths longer than 7.5 Å, then a pronounced matrix effect can be expected and

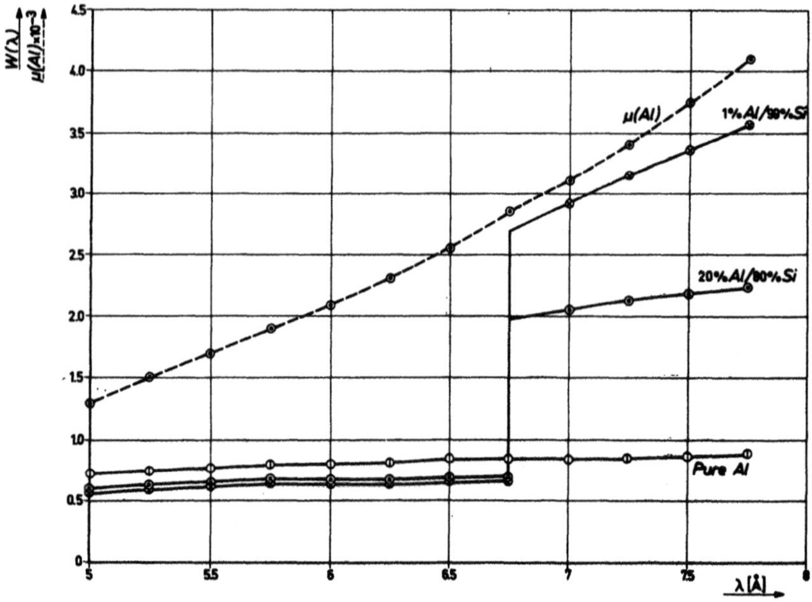

Fig. 22

non-linear calibration lines will result. These effects have to be considered when choosing an effective wavelength for correction calculations.

QUESTION 35

Quantitative diffractometry - determination of Anatase in Rutile

A A sample is known to contain two crystalline modifications of TiO_2, namely rutile and anatase. Spectrographic analysis failed to reveal the presence of other elements, so the sample can be assumed to consist only of TiO_2. A scan is made with the diffractometer and integrated intensities measured for the (110) reflection of rutile (I_R) and for the (101) reflection of anatase (I_A). Background measurements were made from a suitable part of the background. The measured intensities are as follows:

	peak + background (cts)	background (cts)
I_R	80,892	7,745
I_A	3,942	3,458

The following relationship is known to exist between the concentration of rutile (C_R) and the concentration of anatase (C_A):

$$\frac{C_R}{C_A} = K\frac{I_R}{I_A} \qquad (21)$$

where K is a proportionality constant. Previous experiments 9) indicated a value of 1.33 for K.
Calculate the concentration of anatase.

B The concentration of anatase determined as above may be checked using the "spiking method". A known weight of anatase (150 mg) was added to 2.16 grams of the original sample and the intensities for anatase peak and background re-measured. The following data were obtained:

	peak + background (cts)	background (cts)
I_A	10,645	4,382

What is the concentration of anatase determined by these data?

ANSWER 35

A An expression is given relating the concentration and intensity:

$$\frac{C_R}{C_A} = K\frac{I_R}{I_A}$$

The sample consists only of TiO_2, so it follows that $C_R + C_A = 100\%$.

Substitution of C_R by $(100-C_A)$ gives

$$C_A = 100\frac{1}{1+K(I_R/I_A)}$$

from the data given the peak counting rates corrected for background are

$$I_R = 80,892 - 7,745 = 73,147 \text{ cts}$$

$$I_A = 3,942 - 3,458 = 484 \text{ cts}$$

Thus:

$$C_A = 100 \cdot \frac{1}{1 + 1.33\ (73,147/484)} = 0.51\%$$

B In the spiking method the peak-background counting rate on the anatase peak will increase proportionally with the amount of extra anatase added. Thus

$$\frac{I_c}{I_{c+x}} = \frac{c}{c+x}$$

where I_c is the intensity due to the unknown concentration c of anatase in the sample, and I_{c+x} is the intensity after adding the extra anatase x.

The percentage of anatase added is simply

$$\frac{0.15 \times 100}{2.16 + 0.15} = 6.49\ \%$$

However in adding the anatase the original concentration changes by

$$2.16/2.31 = 0.935\ c$$

I_c is already known from the previous part of the excercise, i.e. 484 cts.

I_{c+x} can be calculated from the data given: 10,645 − 4,382 = 6,263 cts.

Substitution of these values in Equation (21) gives:

$$\frac{484}{6,263} = \frac{c}{0.935\,c + 6.49}$$

from which $c = 0.54\%$

QUESTION 36

Quantitative X-ray diffractometry use of internal standard

The weight fraction of α quartz (SiO_2) has to be determined in a natural specimen. To correct for the unknown absorption in this specimen, 200 mg KCl are added to 1000 mg of the sample, both very finely ground, and well mixed. Determine the weight fraction of α quartz, using the following data:

	pure material	in sample after adding KCl
intensity of α quartz line (cts)	150,000	4720
intensity of reference KCl line (cts)	100,000	20000

The values of the mass absorption coefficients for Cu Kα can be taken as:

$$K = 150; Cl = 106; Si = 63; O = 12$$

The KCl line is chosen such that the d-spacing is close to the d-spacing of the quartz line, but also such that both are well resolved. Effects of particle size and preferred orientation can be ignored, the filling of all specimens can be taken to be ideal. The experimental conditions were such that direct comparison of the intensities given is possible.

ANSWER 36

The ratio of the intensity of a certain diffraction line of a given compound in a mixture and in a sample, is given by:

$$\frac{I_m^a}{I_0^a} = c_a \frac{\mu_0^a}{\mu_m} \qquad (22)$$

where I_m^a and I_0^a are the intensities measured in the sample and in the pure compound respectively, c_a is the weight fraction of compound a in the sample, μ_0 and μ_m are the mass absorption coefficients for Cu Kα for the pure substance a and the sample.

For the reference line a similar equation can be set up:

$$\frac{I_m^b}{I_0^b} = c_b \frac{\mu_0^b}{\mu_m}$$

If this is applied to the sample after adding KCl, μ_m can be eliminated giving the following equation:

$$c_a' = c_b \cdot \frac{I_0^b}{I_0^a} \cdot \frac{\mu_0^b}{\mu_0^a} \cdot \frac{I_m^a}{I_m^b}$$

where c_a' applies to the concentration after adding KCl. Thus in the original sample the concentration was:

$$c_a = \frac{1200}{1000} \cdot \frac{200}{1200} \cdot \frac{100000}{150000} \cdot \frac{4720}{20000} \cdot \frac{\mu_o^b}{\mu_o^a} = 0.0315 \frac{\mu_o^b}{\mu_o^a}$$

From the absorption data given it can be calculated that for pure quartz $\mu_0^a = 36$ and for pure KCl $\mu_0^b = 129,5$.

Substitution gives a value of $c_a = 0.117$ or 11.7%.

QUESTION 37

Estimation of expected intensities in mixtures

The intensity of the 3.34 A line of a quartz (α SiO_2) has to be estimated in a mixture consisting of 10% α quartz and 90% $CaSiO_3$. Nickel filtered Cu Kα radiation is used. The intensity of the 3.34 Å line in pure α quartz was 100 c/s, measured under the same conditions. What is the expected intensity from the mixture.

The following mass absorption coefficients for Cu Kα radiation can be used:

	(μ) Ca	(μ) Si	(μ) O
Cu Kα	180	63	12

ANSWER 37

The intensity of an X-ray line is approximately inversely proportional to the mass absorption coefficient of the matrix for the radiation in question, in this case Cu Kα. Thus in general terms:

$$\frac{(c/s/\%) \text{ of a line in mixture A}}{(c/s/\%) \text{ for the same line in mixture B}} = \frac{(\mu) \text{ mixture B}}{(\mu) \text{ mixture A}}$$

It is stated that pure α quartz (mixture B) gives 100 c/s for the 3.34 Å line, i.e. 1 c/s/%. By dividing the value by the μ for (10% α quartz +90% $CaSiO_3$) mixture B and multiplying by the μ for pure α quartz, the c/s/% for the mixture can be obtained.

First the μ values for α quartz and the mixture must be calculated. The general expression for the absorption coefficient of a matrix is:

$$\mu_{matrix} = \sum_i (\mu_i w_i)$$

where μ_i and w_i are individual mass absorption coefficients and weight fractions.

The calculation of the absorption coefficient for a mixture is given in Fig. 23. The mixture will give $[1 \times \frac{35.9}{77.7}]$ c/s/% = 0.462 and 10% of SiO_2 in the mixture will give (10 x 0.462) = 4.62 c/s.

FIGURE 23

$$\mu_{matrix} = \sum_i (\mu_i w_i)$$

		SiO_2			$CaSiO_3$	
		Si	O2	Ca	Si	O3
$\mu(SiO_2)$	=	(63 x 0.468) +	(12 x 0.532)			
	=	29.5 +	6.4			
	=	35.9				
μ (mixture)	=	0.1[(63 x 0.468) +	(12 x 0.532)] +	0.9[(180 x 0.345) +	(63 x 0.242) +	(12 x 0.413)]
	=	0.1 (29.5 +	6.4) +	0.9 (62.1 +	15.2 +	15.0)
	=	3.59	+		74.1	
	=		77.7			

QUESTION 38

Quantitative x-ray diffractometry

Very often the quantity of a compound has to be determined in a complex mineral, the composition and absorption coefficients of which are unknown. To correct for these unknown facts, the attenuation of a diffraction line arising from the bottom of the sample holder may be used. The following intensities were collected in measuring the percentage quartz in a mineral with a conventional diffractometer, utilizing a sample holder with a nickel bottom.

I_0^{Ni} for a nickel line for the empty sample holder = 1,000,000 cts

I_s^{Ni} for the same nickel line for the sample holder filled with pure quartz = 150,000 cts

I_m^{Ni} for the same nickel line for the sample holder filled with the mineral = 25,000 cts

I_s^q for a quartz line for pure quartz = 100,000 cts

I_m^q for the same quartz line for the mineral = 6,500 cts

The experimental conditions were the same for measurements of the Ni-line but different for both the quartz line measurements.

A Determine the volume fraction of the quartz in the mineral. It can be assumed that the filling factor and particle size distribution are the same for the quartz and the mineral sample.

B Determine the weight fractions if the specific gravities (s.g.) are found to be: s.g. quartz = 2.64; s.g. mineral = 2.8. Can any of the given measurements be dispensed with. If so under what assumptions.

ANSWER 38

A Let the volume fraction of quartz in the mineral be called v_q. The effective volume which contributes to the intensity of the quartz line is proportional to va, and also to the depth of penetration of the radiation used. This penetration depth is inversely proportional to μ_l, the linear absorption coefficient. The ratio of the intensities measured for the quartz line will thus be a function of the ratio of the linear absorption coefficients, if the proportionality constants are taken to be the same in both cases, as the irradiated area is the same in both cases.

$$\frac{I_m^q}{I_s^q} = \frac{v_q \cdot v_m}{v_s} = \frac{v_q \cdot v_s}{\mu_m} \quad \text{or} \quad v_q = \frac{I_m^q \cdot \mu_m}{I_s^q \cdot \mu_s}$$

If the filling factor and the density were known, μ_s could be calculated from tabulated mass absorption data.

However, if these data are not known, the measured intensities for for the nickel line give the following equations:

$$I_s^{Ni} = I_0^{Ni} \exp(-\mu_m b_s)$$

where b_s is the path length for the radiation through the sample for this Ni diffraction line, this path lenght is the same as in the mineral, assuming the same filling factor.

$$I_m^{Ni} = I_0^{Ni} \exp(-\mu_m b_s), \text{ or}$$

$$\ln \frac{I_0^{Ni}}{I_m^{Ni}} = \mu_m b_s \quad \text{and} \ln \frac{I_0^{Ni}}{I_s^{Ni}} = \mu_s b_s \quad \text{or}$$

$$\frac{\mu_m}{\mu_s} = \frac{\ln I_0^{Ni}/I_m^{Ni}}{\ln I_0^{Ni}/I_s^{Ni}}, \quad \text{or } v_q = \frac{I_m^q \log I_0^{Ni}/I_m^{Ni}}{I_s^q \log I_0^{Ni}/I_s^{Ni}}$$

Filling in the measured data gives:

$$v_q = \frac{6500 \times \log \frac{1000}{25}}{100000 \times \log \frac{100}{15}} = 0.065 \times 1.95 \approx \underline{12,7\%}$$

B The weight fraction is given by:

$$12.7 \times \frac{2.64}{2.8} = \underline{12.0\%}$$

The layer has to be very thin to measure the Ni-line after absorption but not too thin to reach the effective depth for the quartz line.

If the densities and the effective thickness of the samples are known, then the procedure may be simplified. This may happen when the sample holder is filled to a certain given level, and the filling factor may be taken as unity, e.g. after pressing.
The path length for the nickel diffraction line may then be calculated from the known diffraction angle.
In that case the measurement of the Ni-line for the empty holder is unnecessary.

$$\frac{I_m^{Ni}}{I_s^{Ni}} = \exp-\left((\mu_m^x f_m b_m) - (\mu_s^x f_s b_s)\right)$$

In this equation I_m^{Ni}, I_s^{Ni}, f_m can be measured. The μ's represent mass absorption coefficients. μ_s^x, f_s, b_s can be calculated or found in tables, thus μ_m^x can be derived.

This enables the weight fraction to be found according to:

$$c_q = \frac{I_m^q \cdot \mu_m^x}{I_s^q \cdot \mu_s^x}$$

QUESTION 39

Determination of dead time

It is necessary to determine the dead time of an X-ray spectrometer using the $K\beta/K\alpha$ counting ratio method. The following data were obtained on the $K\alpha$ and $K\beta$ lines of tin by varying the X-ray tube current and keeping the kV constant.

measurement	$K\alpha$ intensity (c/s)	$K\beta$ intensity (c/s)
1	2,500	833
2	5,000	1,670
3	7,500	2,515
4	10,000	3,370
5	25,000	8,555
6	50,000	17,550
7	75,000	27,100
8	100,000	37,500

The true counting rate (R_T) is related to the measured counting rate (R) and the dead time (τ) as follows:

$$R_T = \frac{R}{1 - R\tau} \tag{9}$$

Determine τ from these data. The true ratio may be taken from the first measurement, 2500 : 833 = 3

ANSWER 39

If the true counting rate is changed by varying the current on the X-ray tube, then the ratios the intensities of a Kα and the corresponding Kβ line should be constant.

The true counting rate $R_T = \dfrac{R}{1 - R\tau}$ where τ is the dead time.

By taking two widely different values, e.g. measurements 1 and 8 from the given data, and by using the given formula twice for R_α and twice for R_β an equation may be set up, with only τ as the unknown. This calculation is rather inaccurate, just using one set of data and a graphical solution gives in general a more reliable result. If the true ratio C is known, the calculation can be simplified.

Call the ratio of the true $R_\alpha : R_\beta = C$, then

$$R_\alpha^T = C R_\beta^T$$

or

$$\frac{R_\alpha}{1 - R_\alpha \tau} = \frac{C \cdot R_\beta}{1 - R_\beta \tau}$$

or

$$R_\alpha (1 - \tau R_\beta) = C \cdot R_\beta (1 - \tau R_\alpha)$$

or by dividing through by $R_\alpha R_\beta$ and rearranging:

or

$$\frac{C}{R_\alpha} - \frac{1}{R_\beta} = (C - 1)\tau$$

$$(C \cdot R_\beta - R_\alpha) = (C - 1) R_\alpha R_\beta \tau$$

The equation can be used to calculate τ for any set of measurements i.e. 1 - 8 if C is known. The mean value of these seven calculations would give τ. Due to counting statistics the lower values would not give accurate results.

A plot of the expression is shown in Fig. 23 and the data are collected in an approximate form in Table 1. From the slope of the curve it can be found that $\underline{\tau = 1.6\ \mu s.}$

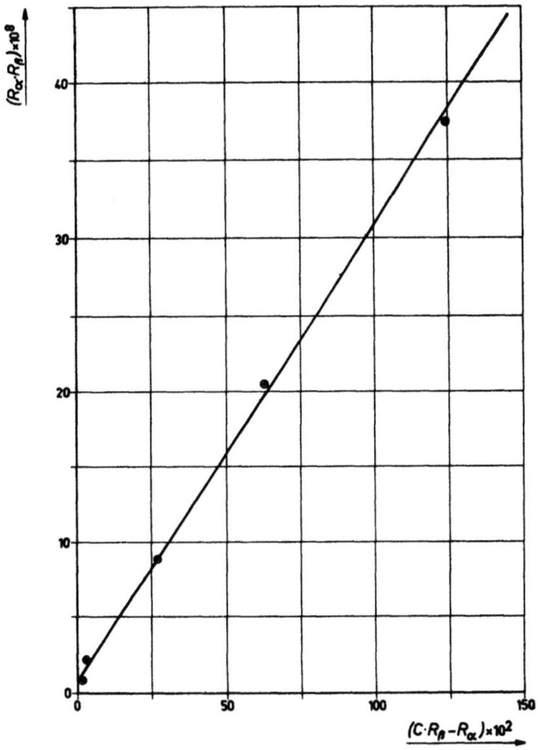

Fig. 23

TABLE 1

R_α	R_β	$R_\alpha . R_\beta . 10^{-6}$	$C \cdot R_\beta - R_\alpha$
5000	1670	8.4	10
7500	2515	19	45
10000	3370	34	110
25000	8555	214	665
50000	17550	880	2650
75000	27100	2030	6300
100000	37500	3750	12500

QUESTION 40

Estimation of gold plating thickness

The following intensities were obtained from a series of gold plated steel samples measured on the Au Lα line.

sample	scaler value (cts)	gold thickness (μm)
1	164.548	0.5
2	322.926	1.0
3	524.319	2.0
4	737.634	4.0
5	832.411	6.0
6	869.450	8.0

Measuring conditions were as follows:
tungsten target tube, 48 kV - 20 mA; LiF crystal; scintillation counter; counting time 20 s.

Prepare a calibration curve of scaler value against gold thickness and estimate the maximum thickness that could be determined, assuming a 3σ confidence level for the counting process.

It can be assumed that further increase in the gold thickness beyond 8 μm causes no significant increase in the counting rate.

ANSWER 40

A calibration curve is plotted of scaler value (ordinate) against gold thickness (abscissa) as shown in Fig.24.

It is stated that increase in the gold thickness beyond 8 μm causes no significant increase in counting rate. The number of counts obtained on sample 6 can then be used to calculate the statistical limits of the number of counts.

$$\sigma = \sqrt{N} \text{ and } 3\sigma = 3\sqrt{N}$$

for sample 6, $N = 869.450 \therefore 3\sigma = 3\sqrt{869.450} = 2,797$

These limits i.e. $869.450 \pm 2,797$ (3σ), can now be plotted as in Fig.24. Where the lower limit cuts the calibration curve is a point representing the maximum gold thickness determinable.

As seen from fig.24 this point occurs at 7.75 μm.

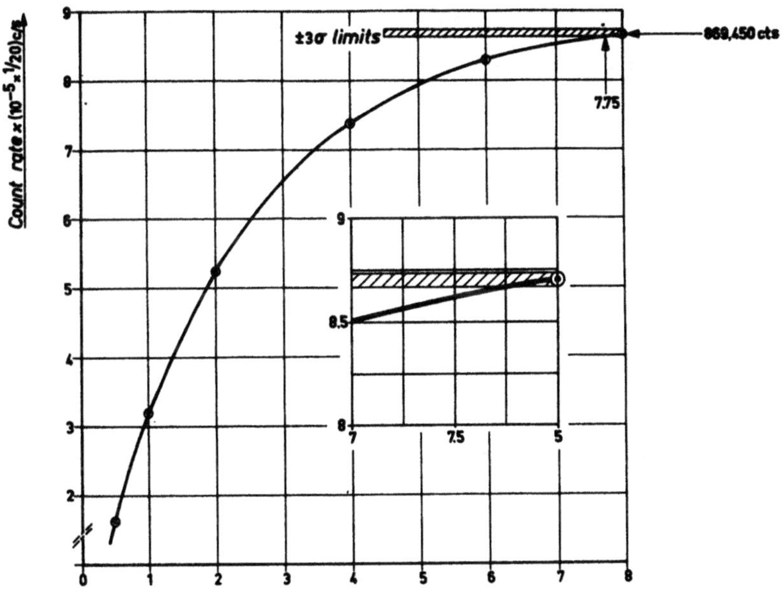

Fig.24

QUESTION 41

Measurement of coating thickness on irregular shaped objects

The thickness of Cr plating on a piece of iron of rather irregular shape has to be measured.

The following intensities were measured with a conventional spectrometer with a take-off angle of 30°:

$$\text{Fe K}\alpha = 530 \text{ c/s} \qquad \text{Fe K}\beta = 260 \text{ c/s}$$

A flat disc of similar iron without plating gave:

$$\text{Fe K}\alpha = 12500 \text{ c/s} \qquad \text{Fe K}\beta = 2250 \text{ c/s}$$

under the same experimental conditions.

Find the thickness of this Cr coating.

The mass absorption coefficients for Cr can be taken as:

$$\text{Fe K}\alpha = 460 \text{ cm}^2/\text{g} \qquad \text{Fe K}\beta = 370 \text{ cm}^2/\text{g}$$

Density for Cr = 7.0 g/cm^3.

ANSWER 41

The Fe Kα radiation excited in the iron is absorbed in the Cr layer according to $I = I_0 \exp.(-\mu \rho b)$ where b is the path length. Unfortunately I_0, the unabsorbed intensity of Fe-radiation, is not known for the irregular shaped piece of iron without a plating, because the measured intensity is dependent upon the area irradiated. The ratio $I\alpha / I\beta$, however, is independent of shape or area. The Kβ radiation is less absorbed in the Cr layer. The ratio of the Kα and Kβ intensity will thus be a function of the Cr thickness.

$$I^{\alpha} = I_0^{\alpha} \exp(-\mu \rho b) \qquad (24)$$
$$= I_0^{\alpha} \exp(-460 \times 7.0 \times X \times \operatorname{cosec} 30^{\circ})$$

where X is the average layer thickness

$$\therefore \quad I^{\alpha} = I_0^{\alpha} \exp(-6440\, X)$$

$$\text{similarly} \quad I^{\beta} = I_0^{\beta} \exp(-5180\, X)$$

$$\text{thus} \quad \frac{I^{\alpha}}{I^{\beta}} = \frac{I_0^{\alpha} \exp(-6440\, X)}{I_0^{\beta} \exp(-5180\, X)}$$

$$\text{or} \quad \frac{530}{260} = \frac{12500}{2250} \exp(-1260\, X)$$

$$2.038 = 5.556 \exp(-1260\, X)$$

$$\therefore 2.73 = \exp(1260\, X)$$

$$\underline{X = 8\ \mu\text{m}}$$

This method gives optimum results when the μ's for two wavelengths are very different. This can happen for an α and β radiation on both sides of the absorber. Two wavelengths of different elements can be used only if the chemical composition of the base material and standard is fairly constant. The α:β method is dependent also on the change in matrix absorption, although to a lesser extent.

This method gives a good indication of over-all thickness of the coating. Small holes, scratches, etc. influence the results less than if only the absorption of one line from the base material was measured.

QUESTION 42

Background variation

A scan is being made on a rock sample in the region of the barium K spectra.

The following conditions are being used: tungsten anode X-ray tube at 100 kV - 20 mA; LiF (220) crystal, $2d$ = 2.848 Å. NaI/Tℓ scintillation counter. An automatic pulse height selector is used in which all pulses coming from the detector are attenuated with decrease of wavelength. In effect this means that all pulses arrive at the same voltage level irrespective of the wavelength of the radiation incident upon the detector. A fixed window is employed, the width of which is approximately twice the peak width at half height of Fe Kα radiation. The spectra obtained shows a sudden increase in intensity (about 15%) at an angle of 15.6 °2θ i.e. at 15.2 °2θ the background intensity is about 15% higher than at 14.8 ° 2θ.

Explain the reason for this background variation; should the background for Ba Kα be measured to the low or high angle side of the Kα line.

ANSWER 42

The variation of background is due essentially to an escape peak phenomenon.

Since the average atomic number of a rock sample is relatively low, its scattering power for primary radiation is high, resulting in high background. The scattered radiation is diffracted by the analysing crystal and at low angles (i.e. where n in the Bragg law is generally equal to unity) the recorded background is almost exclusively first order.

An escape peak will be formed where the energy of radiation entering the scintillation counter is in excess of the K absorption edge energy of iodine (33.1 keV). The distribution of pulses from the counter will contain the distributions above an energy of 33.1 keV but only one distribution below this energy. The overall effect is shown in Fig.25, which shows the pulse amplitude distribution obtained at 13 °2θ (i.e. 38.5 keV) and 16°2θ(i.e. 31.3 keV). As anticipated the distribution of pulses contains two maxima at 13°2θ, but only at 16 °2θ.

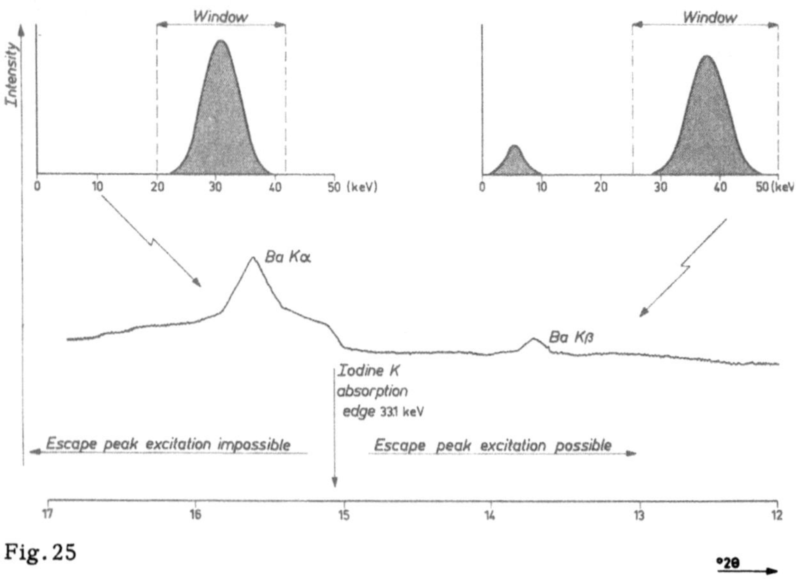

Fig. 25

However, it is immediately apparent that the escape peak distribution falls outside the window setting* and since the total counting rate is always proportional to the area under the pulse amplitude distribution, the background intensity for the angle corresponding to energies greater than 33.1 keV is reduced in the proportion of photopeak to photopeak plus escape peak (about 85% to 100%).

It will be apparent that the angle at which the phenomena occurs will be that corresponding to 33.1 keV (≡ 0.374 Å) i.e.

$$\theta = \sin^{-1}\left(\frac{0.374}{2.848}\right) = 7.53°$$

$$2\theta = 15.06°$$

Since the true background position is that above 15.06 ° it is best to measure the background for Ba Kα (2Θ = 15.06 °) to the high angle side.

---

*) The position of the escape peak in the pulse amplitude distribution will be equal to a voltage corresponding to:

$$\frac{\text{volts (photopeak)}}{\text{volts (escape peak)}} = \frac{\text{energy (incident radiation)}}{\text{energy (incident radiation)} - (\text{energy } I \text{ K}\alpha)}$$

i.e. at 13° at a voltage corresponding to (38.5 - 33.1) keV = 5.4 keV.

QUESTION 43

Use of primary filters for removal of characteristic tube lines

It is necessary to measure low concentrations of chromium using a spectrometer equipped with a chromium anode X-ray tube. A titanium filter can be fitted over the tube window to minimize the intensity of the Cr Kα radiation arising from the scattered tube radiation. Two thicknesses of titanium are available, at 0.13 mm and 0.011 mm.

The following data were obtained on a sample containing almost exactly 1% of chromium and a "blank" sample containing no chromium. The Cr Kα line was used.

	thickness	c/s for 1% Cr-sample	c/s for blank sample
foil 1	0.13 mm	391	23.9
foil 2	0.011 mm	7,150	1,520
no foil	0	31,100	24,000

A Calculate the relative usefulness of the two foils for the reduction of the Cr Kα radiation from the X-ray tube compared to the situation when no filter at all is present.

B Calculate the percentage error obtained in an analysis time of 100 s for both foils.

In each case it can be assumed that the method of "optimum time" is used, and the following expression can be employed:

$$\varepsilon\% = \frac{100}{\sqrt{T}} \cdot \frac{1}{\sqrt{R_p} - \sqrt{R_b}} \qquad (6)$$

where T is the total analysis time and R_p and R_b are the peak and background counting rates respectively. ε is the percentage error.

ANSWER 43

A In order to estimate the relative efficiency of different sets of equipment parameters it is necessary to use a "figure of merit". From the given equation it will be apparent that $\varepsilon\%$ will be at a minimum when $\sqrt{R_p} - \sqrt{R_b}$ is at a maximum, provided that the analysis time T is constant. Thus $\sqrt{R_p} - \sqrt{R_b}$ can be taken as the figure of merit (F.O.M.).

In the given example:

foil	R_p	R_b	$\sqrt{R_p}$	$\sqrt{R_b}$	F.O.M. = $\sqrt{R_p} - \sqrt{R_b}$
1	391	23.9	19.8	4.9	14.9
2	7150	1520	84.6	39.0	45.6
none	31100	24000	176.4	154.9	21.5

Since foil 2 gives by far the largest figure of merit this represents the optimum choice.

B Equation 6 can also be used to calculate the actual error $\varepsilon\%$. Replacing $\sqrt{R_p} - \sqrt{R_b}$ by F.O.M. gives:

$$\varepsilon\% = \frac{100}{\sqrt{T}} \cdot \frac{1}{\text{F.O.M.}}$$

As the analysis time is given as 100 s.

$$\varepsilon\% = \frac{10}{\text{F.O.M.}}$$

Thus for foil 1 $\quad \varepsilon\% = \frac{10}{14.9} = \underline{0.67\%}$

and for foil 2 $\quad \varepsilon\% = \frac{10}{45.6} = \underline{0.22\%}$

QUESTION 44

Calculation of ß filter thickness and transmission

A Ni filter has been constructed in such a way that only 2% of the incident Cu Kβ radiation may pass. Given that the density of Ni is 8.92 g/cm^3 and that the mass absorption of Ni for Cu Kα and Cu Kβ is 49.2 cm^2/g and 286 cm^2/g respectively, calculate:

A The thickness of the filter

B The percentage of incident Cu Kα radiation that can pass the filter.

ANSWER 44

When a parallel beam of X-rays strikes the surface of an absorber at normal incidence, the intensity of the transmitted beam can be calculated from the Beer-Lambert law.

$$I = I_0 \exp(-\mu \varrho x) \qquad (24)$$

where x is the thickness of the absorber, ϱ its density and μ its mass absorption coefficient for the measured radiation. Rearranging Equation (24) in $\log_{10}$ form:

$$\log \left[\frac{I_0}{I}\right] = 0.434\, \mu \varrho x \qquad (24')$$

a) In the given example for Cu Kβ (I_0/I) is 100/2 or 50. Substituting in the above equation gives:

$$\log 50 = 0.434 \times 286 \times 8.92 \times x$$

$$\text{or } \underline{x = 15.2\ \mu m}$$

b) Equation (24') can also be used to calculate the percentage of Cu Kα transmitted. Thus:

$$\log \left[\frac{100}{I}\right] = 0.434 \times 49.2 \times 8.92 \times 0.00152$$

$$\log 100 - \log I = 0.289$$

$$\text{or } \log I = 2 - 0.289$$

$$\therefore\ I = \text{antilog}(1.811)$$

$$= 51.4\%$$

Thus the transmission of the given filter for Cu Kα radiation is 51.4%

QUESTION 45

Utilization of the ß filter and pulse height selector in diffractometry

A filter is very often used in X-ray diffractometry to reduce the background level and the intensity of β-radiation. This filter can be placed either between X-ray tube and specimen or between the specimen and detector.

A Which position gives the best result in general and when should the alternative position be used.

B Can the pulse height selector be used to eliminate the background and the β-radiation arising with a diffractometer using Cu K radiation, 50 kV - 20 mA, and a Xe-filled proportional counter.

ANSWER 45

A When using copper radiation a Ni-filter is employed to eliminate the β-radiation and to reduce the background. It will absorb white radiation by a factor which is dependent upon its thickness (c.f. Q44 to calculate the optimum thickness). It does not make any difference if this filter is placed before or after the specimen as far as the elimination of the coherent scattered primary or β-radiation is concerned. However, the atoms in the specimen may be excited by the primary X-rays and can then emit fluorescence radiation, which increases the general background level. This sample fluorescence is considerably reduced when the filter is placed in front of the detector. It is thus in general advisable to place the filter between specimen and detector. Positioning the filter in front of the receiving slit assembly has the additional advantage that most of the excited radiation within the filter does not enter the detector.

However, there is an exception to this general rule. If investigation is being made of, for instance, a Ni-specimen with Cu K radiation, the Ni atoms can be excited by the Cu Kβ radiation. The resulting Ni K radiation can pass the Ni filter and give rise to a general high background. In this case, the filter should be placed in front of the specimen to eliminate the Cu Kβ radiation before it can excite Ni fluorescence radiation.

A general rule could then be formulated: "The β-filter should always be placed between specimen and detector, unless the specimen contains the filter element in major concentration".

B The absorption of the filter is not very efficient for very short wavelengths or for wavelengths slightly longer than the Kα radiation. The proportional counter in combination with pulse height selection transmits the Kα radiation fully, whereas the filter generally absorbs about half of the Kα radiation. The proportional counter - pulse height selector combination thus shows great advantage. It can reduce the intensity of neighbouring softer radiation to a large extent (c.f. Q14). However, it cannot significantly reduce the Kβ radiation. For example, using Cu K radiation: Kα = 8.0 keV

Kβ = 8.9 keV

Assuming a detector resolution (i.e. peak width at half height) of 18% for the detector for Cu Kα radiation, the window of the pulse height selector should be set to approximately 45% of 8.0 keV = 3.6 keV to prevent radiation between 6.2 and 9.8 keV passing the discriminator. The larger part of the β-radiation will thus be registered, but the white radiation above about (8 + 2 x 1.8) = 11.6 keV cannot contribute to the background radiation.

It is thus clear that for recording simple diffractograms where the β-maxima will not coincide with the α-maxima, a proportional counter with pulse height selector gives better results than a β-filter; but in general, a β-filter must be employed to improve the overall resolving power.

QUESTION 46

Determination of lattice constants for a cubic lattice

The following 12 lines were obtained from a crystalline powder. Data were obtained using nickel filtered Cu Kα radiation. The powder is known to belong to the cubic system.

line	d (Å)	relative intensity
1	3.157	94
2	1.931	100
3	1.647	35
4	1.366	12
5	1.253	10
6	1.1150	16
7	1.0512	7
8	0.9657	5
9	0.9233	7
10	0.9105	1
11	0.8637	9
12	0.8330	3

Index the lines in terms of their Miller indices (hkl) and calculate the lattice constant of the cubic lattice. Establish the type of cubic lattice.

ANSWER 46

The following relationship holds for a cubic lattice:

$$d_{hkl} = \frac{a_0}{\sqrt{h^2+k^2+l^2}} \qquad (4)$$

where a_0 is the lattice constant.

It follows, therefore, that $(h^2+k^2+l^2)$ is inversely proportional to $(d_{hkl})^2$. Indexing is achieved by calculating $1/d^2$ for each line and looking for a constant factor in the values that will yield whole numbers for $(h^2+k^2+l^2)$. Certain values, e.g. 7, 15, 23, must be missing as they cannot be obtained from the sum of 3 integers squared. In fact the constant term established above will be the square of the cell constant (a_0), i.e. from Equation (4)

$$(h^2+k^2+l^2) = \frac{(a_0)^2}{d^2}$$

line	d	d^2	$1/d^2$	$(h^2+k^2+l^2)$	hkl
1	3.157	9.967	0.1003	3	(111)
2	1.931	3.729	0.2682	8	(220)
3	1.647	2.713	0.3686	11	(311)
4	1.366	1.866	0.5358	16	(400)
5	1.253	1.570	0.6369	19	(331)
6	1.1150	1.243	0.8045	24	(422)
7	1.0512	1.105	0.9053	27	(511) (333)
8	0.9657	0.9326	1.072	32	(440)
9	0.9233	0.8525	1.173	35	(531)
10	0.9105	0.8290	1.206	36	(600)
11	0.8637	0.7460	1.340	40	(620)
12	0.8330	0.6939	1.441	43	(533)

A graphical plot (see Fig.26) gives the best average value for a_0 = 5.436 Å. This material crystalizes with the face centred cubic type of lattice.

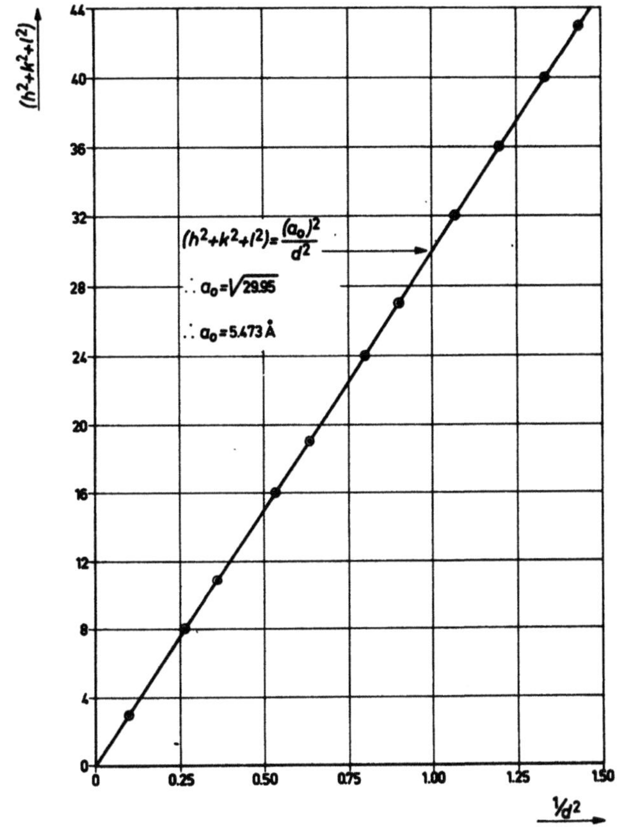

$(h^2+k^2+l^2)$
44
40
36
32
28
24
20
16
12
8
4
0
$(h^2+k^2+l^2) = \frac{(a_0)^2}{d^2}$
$\therefore a_0 = \sqrt{29.95}$
$\therefore a_0 = 5.473$ Å
0
0.25
0.50
0.75
1.00
1.25
1.50
$1/d^2$

Fig. 26

QUESTION 47

Line broadening

Measurements of the nickel (111) reflection are being made on a series of Raney Nickel catalysts. Severe line broadening problems are encountered due to the very small particle size of the powders (about 50 - 100 Å). An estimate is required of the magnitude of the line broadening.

Construct a calibration curve of particle dimension (abscisca) versus line breadth (*B*) in °2θ (ordinate) over the range 10 – 10,000 μm. Use the Scherrer formula to calculate the particle dimension (D_{111}), i.e.

$$D_{111} = \frac{K\lambda}{\beta \cos\theta} \tag{25}$$

Where β (line broadening due to crystallite dimension) = $B \div b$, b being the instrumental broadening. The (111) reflection of nickel occurs at about 44 °2θ when using Cu Kα (λ = 1.542 Å). A value of 0.9 may be taken for the shape factor K.

Measurement on the (111) nickel line from a standard (and unstrained) sample gave a value of 0.33 °2θ for the instrumental broadening (b). Thus

$$\text{line breadth } (B) = \beta + 0.33$$

ANSWER 47

It is stated that the particle dimension D_{111} is given by:

$$D_{111} = \frac{K\lambda}{\beta \cos\theta}$$

and that $\beta = B - b$. Values of K, λ, b and 2θ are given and it is necessary to construct a calibration curve of D_{111} versus B.

Substituting the values given

$$D_{111} = \frac{0.9 \times 1.542 \times 57.3}{(B \times 0.33) \times 0.926} \, (\text{Å})$$

where the factor 57.3 is used to convert the value of β from degrees to radians.

Thus $$B = \frac{84.9}{D} + 0.33$$

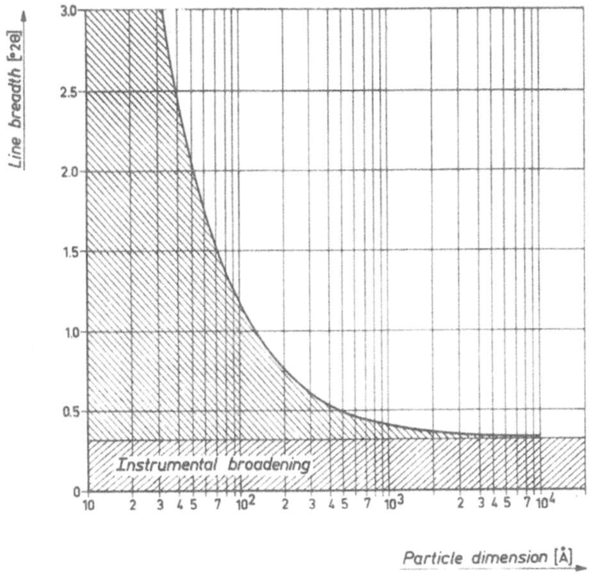

Fig. 27

Fig.27 was constructed from the following table:

D	$84.9/D$	$(84.9/D) + 0.33$
30	2.83	3.16
50	1.70	2.03
100	0.849	1.179
500	0.170	0.500
1000	0.0849	0.415
5000	0.017	0.347
10000	0.0085	0.339

In order to give a rough indication of the effect on the line profile Fig.28 shows calculated line profiles for 50, 100, 1000 and 10.000 Å.

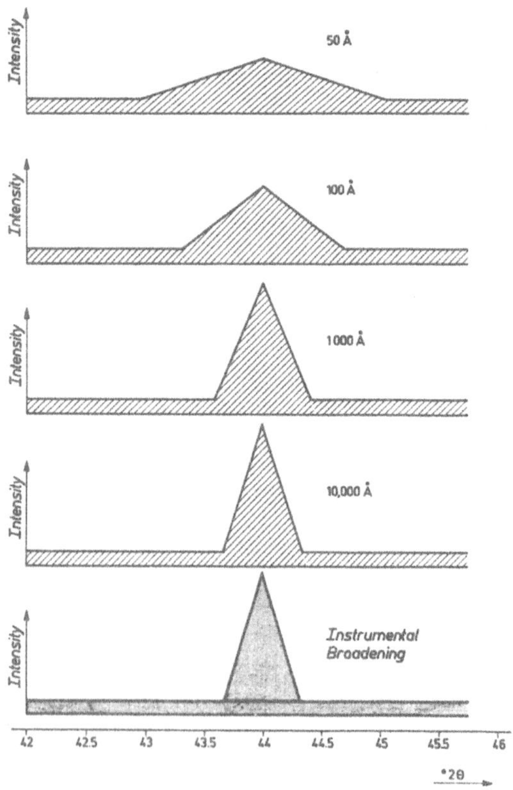

Fig.28

REFERENCES

1) R. Jenkins and J.L. de Vries, "Practical X-ray Spectrometry", Philips Technical Library, London, Macmillan and Co Ltd. (1970)

2) H.P. Klug and L.E. Alexander, "X-ray Diffraction Procedures", New York: Wiley (1954), p. 530.

3) W. Parrish "X-ray Analysis Papers", Philips Technical Library, London, Macmillan and Co. (1965).

4) Kramers, H.A., Phil. Mag., 46, 636, (1923).

5) Spencer, L.V., "Energy dissapation by fast electrons", N.B.S. Monograph No.1, (1959).

6) Compton and Allison "X-rays in Theory and Experiment", New York: Van Nostrand 1935.

7) Jenkins, R., Philips Scientific Reports, FS6,: Philips Scientific Group (1968).

8) Trost, A., Z.f. angew. Physik, 1, (1955), 470.

9) "The determination of anatase in rutile by X-ray diffraction", Philips Scientific Reports, RD2,: Philips Scientific Group (1967).

Table I. Principal Emission Lines of X-ray Spectra (Emission Wavelengths in Å Units)

K Series											
Line		α^*	α_1	α_2	β_1	β_3	β_2	β_4	β_5	—	*K Absorption Edge*
Transition K⟶			L_{III}	L_{II}	M_{III}	M_{II}	N_{II},N_{III}	N_{IV},N_{V}	M_{IV},M_{V}	$O_{II,III}$	
Approximate Intensity (Rel. $K\alpha_1$)		150	100	50	15		5	<1	<1	<1	
Li	3	240.									226.953
Be	4	113.									
B	5	67.									
C	6	44.									43.767
N	7	31.603									31.052
O	8	23.707									23.367
F	9	18.307									
Na	11	11.909			11.617						
Mg	12	9.889			9.558						9.512
Al	13	8.339	8.338	8.341	7.981						7.951
Si	14	7.126	7.125	7.127	6.769						6.744
P	15	6.155			5.804						5.787
S	16	5.373	5.372	5.375	5.032						5.018
Cl	17	4.729	4.728	4.731	4.403						4.397
A	18	4.192	4.191	4.194							3.871
K	19	3.744	3.742	3.745	3.454				3.442		3.437
Ca	20	3.360	3.359	3.362	3.089				3.074		3.070
Sc	21	3.032	3.031	3.034	2.780				2.764		2.758
Ti	22	2.750	2.749	2.753	2.514				2.498		2.497
V	23	2.505	2.503	2.507	2.285				2.270		2.269
Cr	24	2.291	2.290	2.294	2.085				2.071		2.070
Mn	25	2.103	2.102	2.105	1.910				1.897		1.897
Fe	26	1.937	1.936	1.940	1.757				1.745		1.744
Co	27	[illegible]	1.789	1.793	1.621				1.609		1.608
Ni	28	1.659	1.658	1.661	1.500		1.489		1.489		1.488
Cu	29	1.542	1.540	1.544	1.392	1.393	1.381		1.382		1.381
Zn	30	1.437	1.435	1.439	1.296		1.284		1.285		1.281
Ga	31	1.341	1.340	1.344	1.207	1.208	1.196		1.197		1.195
Ge	32	1.256	1.255	1.258	1.129	1.129	1.117		1.119		1.116
As	33	1.177	1.175	1.179	1.057	1.058	1.045		1.049		1.045
Se	34	1.106	1.105	1.109	0.992	0.993	0.980		0.984		0.980
Br	35	1.041	1.040	1.044	0.933	0.933	0.921		0.926		0.920
Kr	36	0.981	0.980	0.984	0.879	0.879	0.866	0.866	0.871		0.866
Rb	37	0.927	0.926	0.930	0.829	0.830	0.817				0.816
Sr	38	0.877	0.875	0.880	0.783	0.784	0.771				0.770
Y	39	0.831	0.829	0.833	0.740	0.741	0.728	0.727			0.727
Zr	40	0.788	0.786	0.791	0.701	0.702	0.690				0.688
Nb	41	0.748	0.747	0.751	0.665	0.666	0.654				0.653
Mo	42	0.710	0.709	0.713	0.632	0.633	0.621	0.620			0.620
Tc	43	0.674	0.673	0.676	0.602						
Ru	44	0.644	0.643	0.647	0.572	0.573	0.562				0.560

K Series											
Line		α^*	α_1	α_2	β_1	β_3	β_2	β_4	β_5	—	*K Absorption Edge*
Transition K⟶			L_{III}	L_{II}	M_{III}	M_{II}	N_{II},N_{III}	N_{IV},N_{V}	M_{IV},M_{V}	$O_{II,III}$	
Approximate Intensity (Rel. $K\alpha_1$)		150	100	50	15		5	<1	<1	<1	
Rh	45	0.614	0.613	0.617	0.546	0.546	0.535				0.534
Pd	46	0.587	0.585	0.590	0.521	0.521	0.510				0.509
Ag	47	0.561	0.559	0.564	0.497	0.498	0.487				0.486
Cd	48	0.536	0.535	0.539	0.475	0.476	0.465				0.464
In	49	0.514	0.512	0.517	0.455	0.455	0.445				0.444
Sn	50	0.492	0.491	0.495	0.435	0.436	0.426				0.425
Sb	51	0.472	0.470	0.475	0.417	0.418	0.408				0.407
Te	52	0.453	0.451	0.456	[illegible]	0.401	0.391				0.390
I	53	0.435	0.433	0.438	0.384	0.385	0.376				0.374
X	54	0.418	0.416	0.421	0.369		0.360				0.359
Cs	55	0.402	0.401	0.405	0.355	0.355	0.346				0.345
Ba	56	0.387	0.385	0.390	0.341	0.342	0.333				0.332
La	57	0.373	0.371	0.376	0.328	0.329	0.320				0.319
Ce	58	0.359	0.357	0.362	0.316	0.317	0.309				0.307
Pr	59	0.346	0.344	0.349	0.305	0.305	0.297				0.296
Nd	60	0.334	0.332	0.337	0.294	0.294	0.287				0.285
Il	61	0.322	0.321	0.325	0.283						
Sm	62	0.311	0.309	0.314	0.274	0.274	0.267				0.265
Eu	63	0.301	0.299	0.304	0.264	0.265	0.258				0.256
Gd	64	0.291	0.289	0.294	0.255	0.256	0.249				0.246
Tb	65	0.281	0.279	0.284	0.246	0.246	0.239				0.238
Dy	66	0.272	0.270	0.275	0.237	0.238	0.231				0.230
Ho	67	0.263	0.261	0.266							0.223
Er	68	0.255	0.253	0.258	0.222	0.223	0.217				0.215
Tu	69	0.246	0.244	0.250	0.215	0.216					0.209
Yb	70	0.238	0.236	0.241	0.208	0.209	0.203				0.202
Lu	71	0.231	0.229	0.234	0.202	0.203	0.197				0.195
Hf	72	0.224	0.222	0.227	0.195	0.196	0.190				0.189
Ta	73	0.217	0.215	0.220	0.190	0.191	0.185				0.184
W	74	0.211	0.209	0.213	0.184	0.185	0.179				0.178
Re	75	0.204	0.202	0.207	0.179	0.179	0.174				0.173
Os	76	0.198	0.196	0.201	0.173	0.174	0.169				0.168
Ir	77	0.193	0.191	0.196	0.168	0.169	0.164	0.163	0.167	0.163	0.163
Pt	78	0.187	0.185	0.190	0.163	0.164	0.159	0.159	0.162	0.158	0.158
Au	79	0.182	0.180	0.185	0.159	0.160	0.155	0.154	0.158	0.153	0.153
Hg	80										0.149
Tl	81	0.172	0.170	0.175	0.150	0.151					0.144
Pb	82	0.167	0.165	0.170	0.146		0.147	0.141	0.145	0.141	0.141
Bi	83	0.162	0.161	0.165	0.142	0.143	0.138				0.137
Th	90	0.135	0.133	0.138	0.117	0.118	0.114		0.116	0.113	0.113
U	92	0.128	0.126	0.131	0.111	0.112	0.108				0.107

*Kα . . . Unresolved $K\alpha_1$ $K\alpha_2$

Table II. Principal Emission Lines of X-ray Spectra (Emission Wavelengths in Å Units)

L Series

Line	α	α1	α2	β1	β2	β3	β4	β5	β6	β7	β9	β10	β15	β17	γ1	γ2	γ3	γ4	γ5	γ6	γ8	l	η	s	t	L Absorption Edges		
Transition L_I →						M_{III}	M_{II}				M_V	M_{IV}				N_{II}	N_{III}	O_{III}										
L_{II} →				M_{IV}										M_{III}	N_{IV}				N_I	O_{IV}	O_I		M_I			L_I	L_{II}	L_{III}
L_{III} →		M_V	M_{IV}		N_V			$O_{IV,V}$	N_I	O_I			N_{IV}									M_I		M_{III}	M_{II}			
Approximate Intensity (Rel. $L\alpha_1$)	110	100	10	80	60	30	20	30	<1	<1	<1	<1	<1	<1	40	10	10	10	<1	20	<1	30	10	<1	<1			
Cl 17																						67.84	67.25					
A 18																						56.212	56.813					
K 19																						47.835	47.325				42.184	
Ca 20	36.393			36.022																		41.042	40.542				35.200	35.561
Sc 21	31.393			31.072																		35.671	35.200					
Ti 22	27.445			27.074																		31.423	30.942				27.29	
V 23	24.309			23.898																		27.836	27.375					
Cr 24	21.713			21.323																		24.840	24.339			16.7	17.9	20.7
Mn 25	19.489			19.158																		22.315	21.864					
Fe 26	17.602			17.290																		20.201	19.73					
Co 27	16.000			15.698																		18.358	17.86					
Ni 28	14.595			14.308																		16.693	16.304					
Cu 29	13.357			13.079																		15.297	14.940				13.010	13.289
Zn 30		12.282		12.096																		14.081	13.719				11.861	12.130
Ga 31		11.313		11.045																		12.976	12.620					
Ge 32		10.456		10.194																		11.944	11.608					
As 33		9.671		9.414																		11.069	10.732			8.108	9.124	9.367
Se 34		8.990		8.735																		10.293	9.959			7.505	8.417	8.645
Br 35		8.375		8.126																		9.583	9.253					
Kr 36																												
Rb 37		7.318	7.325	7.075		6.788	6.821		6.984							6.045			6.754			8.363	8.042			5.997	6.643	6.864
Sr 38		6.863	6.870	6.623		6.367	6.403		6.519							5.644			6.297			7.836	7.517			5.582	6.172	6.387
Y 39		6.449	6.456	6.211		5.983	6.018		6.094							5.283			5.875			7.356	7.040			5.233	5.756	5.962
Zr 40		6.070	6.077	5.836	5.586	5.632	5.668		5.710						5.384	4.953			5.497			6.918	6.606			4.867	5.378	5.583
Nb 41		5.725	5.732	5.492	5.238	5.310	5.346		5.361						5.036	4.654			5.151			6.517	6.210			4.581		5.223
Mo 42		5.406	5.414	5.176	4.923	5.013	5.048		5.048						4.726	4.380			4.837			6.150	5.847			4.299	4.719	4.913
Tc 43																												
Ru 44		4.846	4.854	4.620	4.372	4.487	4.523		4.487						4.182	3.897			4.288			5.503	5 204				4.179	4.369
Rh 45		4.597	4.605	4.374	4.130	4.253	4.289		4.242						3.944	3.685			4.045			5.217	4.922			3.626	3.942	4.129
Pd 46		4.368	4.376	4.146	3.909	4.034	4.071		4.016		3.792	3.799			3.725	3.489			3.822			4.952	4.660			3.428	3.724	3.908
Ag 47		4.154	4.162	3.935	3.703	3.834	3.870		3.808		3.605	3.611			3.523	3.307			3.616			4.707	4.418			3.254	3.514	3.698
Cd 48		3.956	3.965	3.739	3.514	3.644	3.681		3.614		3.430	3.437			3.336	3.137			3.426			4.480	4.193			3.084	3.326	3.504
In 49		3.752	3.781	3.555	3.339	3.470	3.507		3.436		3.268	3.274			3.162	2.980		2.926	3.249			4.269	3.983			2.926	3.147	3.325

Sn	50		3.600	3.609	3.385	3.175	3.306	3.344		3.270	3.156	3.115	3.121			3.001	2.885		2.778	3.086			4.071	3.789			2.778	2.982	3.156
Sb	51		3.439	3.448	3.226	3.023	3.152	3.190		3.115	3.005	2.973	2.979			2.852	2.695		2.639	2.932			3.888	3.607			2.639	2.830	3.000
Te	52		3.290	3.299	3.077	2.882	3.009	3.046		2.971	2.863	2.839	2.847			2.712	2.567		2.511	2.790			3.716	3.438			2.510	2.687	2.856
I	53		3.148	3.157	2.937	2.751	2.874	2.912		2.837	2.730	2.713	2.720			2.582	2.447		2.391	2.657			3.557	3.280			2.389	2.553	2.719
X	54																										2.274	2.429	2.592
Cs	55		2.892	2.902	2.683	2.511	2.628	2.666		2.593	2.485	2.478	2.492			2.348	2.237	2.233	2.174	2.417			3.267	2.994			2.167	2.314	2.474
Ba	56		2.776	2.785	2.567	2.404	2.516	2.555		2.482	2.382	2.376	2.387			2.242	2.138	2.134	2.075	2.309		2.222	3.135	2.862			2.068	2.204	2.363
La	57		2.665	2.674	2.458	2.303	2.410	2.449		2.379	2.275	2.282	2.290			2.141	2.046	2.041	1.983	2.205			3.006	2.740			1.973	2.103	2.259
Ce	58		2.561	2.570	2.356	2.208	2.311	2.349		2.282	2.180	2.188	2.195			2.048	1.960	1.955	1.899	2.110		2.023	2.892	2.620			1.890		2.164
Pr	59		2.463	2.473	2.259	2.119	2.216	2.255		2.190	2.091	2.100	2.107			1.961	1.879	1.874	1.819	2.020		1.936	2.784	2.512			1.811	1.924	2.077
Nd	60		2.370	2.382	2.166	2.085	2.126	2.166		2.103	2.009	2.016	2.023			1.878	1.801	1.797	1.745	1.935	1.855		2.675	2.409			1.735	1.848	1.995
Il	61		2.283		2.081																								
Sm	62		2.199	2.210	1.998	1.882	1.962	2.000	1.779	1.946	1.856	1.862	1.870			1.726	1.659	1.655	1.606				2.482	2.218			1.598	1.702	1.845
Eu	63		2.120	2.131	1.920	1.812	1.887	1.926		1.875	1.788	1.792	1.800			1.657	1.597	1.591	1.544	1.708		1.632	2.395				1.536	1.626	1.776
Gd	64		2.046	2.057	1.847	1.746	1.815	1.853		1.807	1.723		1.731			1.592	1.534	1.529	1.485				2.312	2.049			1.477	1.561	1.709
Tb	65		1.976	1.986	1.777	1.682	1.747	1.785	1.577	1.742	1.659		1.667			1.530	1.477	1.471	1.427				2.234				1.421	1.501	1.648
Dy	66		1.909	1.920	1.710	1.683	1.681	1.720		1.681	1.599					1.473	1.423	1.417	1.374	1.518			2.158	1.898			1.365	1.438	1.579
Ho	67		1.845	1.856	1.647	1.567	1.619	1.658		1.622						1.417	1.371	1.364	1.323	1.462			2.086	1.886			1.318	1.390	1.535
Er	68		1.785	1.796	1.587	1.514	1.561	1.601		1.567	1.494	1.485	1.494			1.364	1.321	1.315	1.276	1.406			2.019	1.757			1.269	1.339	1.482
Tm	69		1.726	1.738	1.530	1.463	1.505	1.544		1.515						1.316	1.274	1.268		1.355			1.955	1.695			1.222	1.288	1.433
Yb	70		1.672	1.682	1.476	1.416	1.452	1.491	1.387	1.466	1.395	1.384	1.392			1.268	1.228	1.222	1.185	1.307	1.243	1.250	1.894	1.635		1.831	1.181	1.243	1.386
Lu	71		1.619	1.630	1.424	1.370	1.402	1.441	1.342	1.419	1.350	1.336	1.343	1.372		1.222	1.185	1.179	1.143	1.260	1.198	1.204	1.836	1.478		1.776	1.140	1.198	1.342
Hf	72		1.569	1.580	1.374	1.327	1.353	1.392	1.298	1.374	1.306	1.291	1.299	1.328	1.437	1.179	1.144	1.138	1.103	1.215	1.155	1.161	1.782	1.523	1.663	1.723	1.099	1.154	1.298
Ta	73		1.522	1.533	1.327	1.285	1.307	1.346	1.256	1.331	1.264	1.247	1.254	1.287		1.138	1.105	1.099	1.065	1.173	1.114	1.120	1.728	1.471	1.612	1.672	1.061	1.113	1.256
W	74		1.476	1.487	1.282	1.245	1.263	1.302	1.215	1.290	1.224	1.204	1.212	1.247	1.339	1.098	1.068	1.062	1.028	1.132	1.074	1.081	1.678	1.421			1.024	1.074	1.215
Re	75		1.433	1.444	1.238	1.206	1.220	1.260	1.177	1.252	1.186	1.165	1.172	1.208	1.293	1.061	1.032	1.026	.993	1.094	1.037	1.044	1.630	1.374			.990	1.037	1.177
Os	76		1.391	1.402	1.197	1.169	1.179	1.218	1.140	1.213	1.149	1.126	1.133	1.171		1.025	.998	.992	.959	1.057	1.001	1.008	1.585	1.328			.956	1.001	1.140
Ir	77		1.352	1.363	1.158	1.135	1.141	1.179	1.106	1.179	1.115	1.090	1.097	1.137		.991	.966	.959	.928	1.022	.967	.974	1.541	1.285			.923	.967	1.105
Pt	78		1.313	1.325	1.120	1.102	1.104	1.142	1.072	1.143	1.082	1.054	1.062		1.166	.958	.934	.928	.897	.988	.934	.941	1.499	1.243			.893	.934	1.072
Au	79		1.277	1.288	1.083	1.070	1.068	1.106	1.040	1.111	1.050	1.021	1.028	1.072	1.128	.927	.905	.898	.867	.956	.903	.910	1.460	1.202	1.352	1.414	.864	.903	1.040
Hg	80		1.242	1.253	1.049	1.040	1.034	1.072	1.010	1.080	1.019	.986	.996	1.041	1.090	.897	.876	.869	.839	.925	.873	.880	1.422	1.164			.836	.872	1.009
Tl	81		1.207	1.218	1.015	1.010	1.001	1.039	.981	1.050	.990	.957	.964	1.012	1.056	.868	.848	.842	.812	.895	.845	.852	1.385	1.127	1.279	1.342	.808	.844	.979
Pb	82		1.175	1.186	.982	.983	.969	1.007	.953	1.021	.962	.927	.934	.984	1.022	.840	.822	.815		.867	.817	.824	1.350	1.092	1.244	1.308	.782	.815	.950
Bi	83		1.144	1.155	.952	.955	.939	.977	.926	.993	.935	.898	.905	.957	.989	.814	.796	.790	.761	.840	.791	.799	1.317	1.058	1.210		.757	.789	.924
Po	84		1.114	1.125	.922	.929	.909	.948	.900	.967				.931		.788		.765			.765								
Fr	87		1.030		.840	.858										.716													
Ra	88		1.005	1.017	.814	.836	.803	.841	.807	.871	.817	.769	.776	.838	.844	.694	.682	.675	.649	.717	.673	.680	1.167	.908			.644	.670	.808
Th	90		.956	.968	.766	.794	.755	.793	.765	.828	.775	.723	.730			.653	.642	.635	.611	.675	.632	.640	1.115	.855	1.011	1.080	.606	.630	.761
Pa	91		.933	.945	.742	.774	.732	.770	.746	.808	.755	.701	.708			.634	.624	.617	.594	.655	.613		1.091	.830					
U	92		.911	.923	.720	.755	.710	.748	.726	.789	.736	.681	.687			.615	.605	.598	.577	.635	.595	.601	1.067	.806	.964	1.035	.569	.592	.722
Np	93		.899		.698	.735										.597													

Additional material from *Worked Examples in X-Ray Analysis*,
ISBN 978-1-4899-2649-4, is available at http://extras.springer.com

EXTRA
MATERIALS
extras.springer.com

GPSR Compliance

The European Union's (EU) General Product Safety Regulation (GPSR) is a set of rules that requires consumer products to be safe and our obligations to ensure this.

If you have any concerns about our products, you can contact us on ProductSafety@springernature.com

In case Publisher is established outside the EU, the EU authorized representative is:

Springer Nature Customer Service Center GmbH
Europaplatz 3
69115 Heidelberg, Germany

Batch number: 10682427

Printed by Printforce, the Netherlands